Environmental Science and Engineering

The ultimate goal of this series is to contribute to the protection of our environment, which calls for both profound research and the ongoing development of solutions and measurements by experts in the field. Accordingly, the series promotes not only a deeper understanding of environmental processes and the evaluation of management strategies, but also design and technology aimed at improving environmental quality. Books focusing on the former are published in the subseries Environmental Science, those focusing on the latter in the subseries Environmental Engineering.

Paolo Porto · Emil Fulajtar · Lee Kheng Heng
Editors

Using ^{137}Cs Resampling Method to Estimate Mean Soil Erosion Rates for Selected Time Windows

Springer

Editors
Paolo Porto
Dipartimento di Agraria
University Mediterranea of Reggio Calabria
Reggio Calabria, Italy

Kazimierz Wielki University of Bydgoszcz
Bydgoszcz, Poland

Lee Kheng Heng
Joint FAO/IAEA Centre of Nuclear
Techniques in Food and Agriculture
International Atomic Energy Agency
Vienna, Austria

Emil Fulajtar
Joint FAO/IAEA Centre of Nuclear
Techniques in Food and Agriculture
International Atomic Energy Agency
Vienna, Austria

Soil Science and Conservation Research
Institute
National Agriculture and Food Centre
Bratislava, Slovakia

ISSN 1863-5520 ISSN 1863-5539 (electronic)
Environmental Science and Engineering
ISBN 978-3-031-52806-4 ISBN 978-3-031-52807-1 (eBook)
https://doi.org/10.1007/978-3-031-52807-1

This Springer imprint is published by the registered company Springer Nature Switzerland AG
The registered company address is: Gewerbestrasse 11, 6330 Cham, Switzerland

Paper in this product is recyclable.

Preface

This publication provides a guideline for applying the ^{137}Cs resampling method for soil erosion assessment. It is one of the outputs of Coordinated research project (CRP) *"Nuclear Techniques for a Better Understanding of the Impact of Climate Change on Soil Erosion in Upland Agro-ecosystems" (D1.50.17)* launched by International Atomic Energy Agency (IAEA).

The Soil and Water Management and Crop Nutrition (SWMCN) Subprogramme of the Joint FAO/IAEA Centre of Nuclear Techniques in Food and Agriculture supports IAEA Member States and FAO Member Countries using the ^{137}Cs techniques for assessing soil erosion spatial distribution and rates, improving agricultural land management and implementing soil conservation for better food security.

The ^{137}Cs method used for soil erosion assessment has significant advantages over the conventional erosion measurements. It provides information on long-term mean soil erosion rates, allows to gain retrospective data, and to express the erosion spatial distribution. However, it has also some disadvantages. Firstly, the time window to which the ^{137}Cs method refers (i.e. since the ^{137}Cs fallout in 1960s and the time of sampling) is becoming too long. In 1990s when the Cs-method became very popular it was just ca. 30 years, but for recent studies it is already ca. 60 years and that is too much for efficiently studying the land use impacts and testing the efficiency of soil conservation measures. The second problem is the availability of some input parameters for conversion models. Finally, the third limitation is the heterogeneity of ^{137}Cs spatial distribution, its micro-variability and problems with representativeness of reference site.

These limitations can be overcome if ^{137}Cs is resampled. In such case the investigated time window is much shorter, and its length can be chosen by researchers. Gaining data on input parameters is easier for shorter period. The older the past the lowest the probability that data (especially the land use, but also rainfall data) will be available. The ^{137}Cs resampling helps also partially reduce the problem with the ^{137}Cs spatial heterogeneity and reference value because there is possibility to compare directly the ^{137}Cs activities from first and second sampling campaign at studied site, without the need of reference site.

Overcoming these limitations makes the ^{137}Cs resampling method very attractive and it was used already in late 1980s but until presence it is used very rarely (just by few tens of studies). These ^{137}Cs resampling studies used different approaches how to interpret the resampling results and this variability hinders the comparison of results presented in individual studies. Therefore, the resampling method needs some standardization, and this became one of the objectives of CRP launched by IAEA and the results are provided in this publication.

The presented guidelines are based on reviewing the earlier ^{137}Cs resampling studies (carried mainly in Canada, USA and Australia), and on own case studies done in Italy, Spain, China and Morocco. All reviewed studies and own case studies were evaluated, compared and discussed. Based on that there was proposed a protocol for implementing the ^{137}Cs resampling in seven steps. This approach can be done in two modifications: (1) traditional reference approach (TRA) and (2) spatial reference approach (SRA).

These guidelines should assist scientists, land managers, environmentalists and students to better assess the impact of soil erosion events and prepare input information for designing soil conservation programmes and implementing soil conservation measures for supporting the sustainable agricultural land management.

Reggio Calabria, Italy — Paolo Porto
Vienna, Austria — Emil Fulajtar
Vienna, Austria — Lee Kheng Heng

Acknowledgements

Joint FAO/IAEA Centre

Nuclear Techniques in Food and Agriculture

We thank the Joint FAO/IAEA Centre of Nuclear Techniques in Food and Agriculture for funding the Coordinated Research Project D15017 'Nuclear Techniques for a Better Understanding of the Impact of Climate Change on Soil Erosion in Upland Agro-ecosystems' from which this publication is resulting. Further, we wish to thank Dr. Lionel Mabit for his effort spent for preparation of this project and valuable contributions when carrying the project management and Dr. Arsenio Toloza for his great support in providing analytical services for this project.

Contents

1 Introduction 1
P. Porto and E. Fulajtar

2 Geographical Overview of ^{137}Cs Resampling Studies 15
P. Porto, E. Fulajtar, D. E. Walling, G. Callegari, V. Cogliandro, C. La Spada, L. Gaspar, A. Navas, H. Q. Yu, W. X. Liu, A. Chappell, Y. Li, M. Benmansour, M. Moustakim, B. Damnati, R. Moussadek, A. Nouira, N. Amenzou, R. Mrabet, and L. Kheng Heng

3 Validating ^{137}Cs Resampling Approach by Comparing with Conventional Erosion Plot Measurements: An Example of Cultivated Site in Italy 111
P. Porto and E. Fulajtar

4 Step-By-Step Protocol to Apply the ^{137}Cs Resampling Approach to Assess Soil Erosion During Chosen Time Periods 129
P. Porto and E. Fulajtar

Chapter 1
Introduction

P. Porto and E. Fulajtar

Abstract This chapter explains the reason for promoting the ^{137}Cs method, its advantages and disadvantages and the need for its standardization. It starts with brief introductory information on the importance of soil erosion problem for agriculture and food security, the role of ^{137}Cs method in soil erosion assessment and advantages and limitations of this method. Further it provides an overview of already existing ^{137}Cs resampling studies as well as the new case studies done by coauthors of these guidelines. Further it describes the content of individual chapters constituting this book and finally it explains the main results of reviewing the existing resampling studies, i.e. classifying the existing resampling approaches to two categories. There are emphasized four major limitations of ^{137}Cs method. Firstly, the time window to which the ^{137}Cs method refers (i.e. since the ^{137}Cs fallout in 1960s and the time of sampling) is becoming too long. In early 1990s when the ^{137}Cs-method became very popular it was just ca. 30 years, but for recent studies it is already ca. 60 years and that is too much for efficient studying the land use impacts and testing the efficiency of soil conservation measures. The second problem is the selection of conversion models, availability of some input parameters for conversion models and validation of models. Third problem is to distinguish Chernobyl ^{137}Cs from bomb ^{137}Cs in Chernobyl affected areas. Finally, the fourth limitation is the heterogeneity of

P. Porto (✉)
Dipartimento di Agraria, University Mediterranea of Reggio Calabria, 89122 Reggio Calabria, Italy
e-mail: paolo.porto@unirc.it

Faculty of Geographical Sciences, Kazimierz Wielki University, Bydgoszcz, Poland

CNR. Istituto per i Sistemi Agrari e Forestali per il Mediterraneo, Sezione Ecologia e Idrologia Forestale, Rende, Italy

E. Fulajtar
Joint FAO/IAEA Centre of Nuclear Techniques in Food and Agriculture, International Atomic Energy Agency, Vienna, Austria
e-mail: emil.fulajtar@nppc.sk

Soil Science and Conservation Research Institute, National Agricultural and Food Centre, Bratislava, Slovakia

P. Porto et al. (eds.), *Using ^{137}Cs Resampling Method to Estimate Mean Soil Erosion Rates for Selected Time Windows*, Environmental Science and Engineering,
https://doi.org/10.1007/978-3-031-52807-1_1

^{137}Cs spatial distribution, its microvariability and problems with representativeness of reference site. These limitations can be overcome if ^{137}Cs is resampled. In such case the investigated time window is much shorter, and its length can be chosen by researchers. Gaining data on input parameters is easier for shorter period. The older the past the lowest the probability that data (especially the land use, but also rainfall data) will be available. The ^{137}Cs resampling helps also partially reduce the problem with the ^{137}Cs spatial heterogeneity and reference value because there is a possibility to compare directly the ^{137}Cs activities from first and second sampling campaign at studied site, without the need of reference site. Overcoming these limitations makes the ^{137}Cs resampling method very attractive. It was used already in late 1980s but until presence it is used very rarely (just by few tens of study sites were investigated). These ^{137}Cs resampling studies used different approaches how to interpret the resampling results and this variability hinders the comparison of results presented in individual studies. Therefore, the resampling method needs some standardization, and this became one of the objectives of Coordinated research project (CRP) *"Nuclear Techniques for a Better Understanding of the Impact of Climate Change on Soil Erosion in Upland Agro-ecosystems" (D1.50.17)* launched by International Atomic Energy Agency (IAEA) in March 2016 and completed in December 2021. The research activities carried under this project resulted in formulating guidelines for the standardised implementation of ^{137}Cs resampling method. The first task towards any standardisation was to review the existing studies. In total 14 such studies were collected. The introductory chapter provides a table with their overview. Further step was to do own case studies testing the ^{137}Cs resampling. The review found that most studies use two major resampling approaches (1) traditional reference approach (TRA) and (2) spatial reference approach (SRA).

Soil erosion by water is a serious environmental problem in many countries of the world. It is well recognized that its impact may affect the economy of these countries and the damage results from both on-site and off-site effects of soil erosion. One of the main on-site effects is related to the loss of soil productivity and the consequent need to re-integrate the soil resources using different strategies and programs. In addition, even if upland soil loss may enhance soil productivity of the downstream lands, the sedimentation in reservoirs and eutrophication of waterways (some of the off-site effects) cause other damages. Soil erosion costs are therefore difficult to estimate, as they are widely variable from area to area. Pimentel et al. (1995) tried to evaluate the economic impact of soil erosion and estimated that the total cost of damages caused by on- and off-site effects is ca. 44,399,000 US$ each year in the USA only. More recently, these costs have been updated in most European countries (Panagos et al. 2018). Based on the results provided by combining bio-physical and macroeconomic models, Panagos et al. (2018) estimated the annual cost of loss in agricultural productivity to be ca. €1.25 billion in the European Union and Italy seems to be the country that suffers the highest economic impact. This problem requires the attention of scientific community and researchers. They should spend their effort searching for various solutions and develop measures to mitigate erosion impact.

Over the last decades, different direct and indirect techniques have been used to measure and predict soil erosion and sedimentation rates in different areas of the world. Among them, conventional monitoring techniques, such as direct measurements of erosion rates at experimental plots (using total collection of sediment or various devices for runoff and sediment partitioning such as multislot divisors, tipping buckets and Coshocton wheels) or in catchments (using hydrological methods for measuring stream discharge and suspended sediment concentration), volumetric measurements of rills (conventional measurements of rill volumes, or using scanners), geodetic measurements of surface changes (using erosion pins), etc., are expensive, hand labour demanding and time consuming and have uncertain accuracy and representativeness, especially at larger scales and over longer periods. The use of empirical or physically based models has significant limitation in terms of validation. The alternative approaches using fallout radionuclides (FRNs), mainly cesium-137 (^{137}Cs), proved to be very effective to assemble information on spatial patterns and long term rates of erosion and deposition (Mabit et al. 2018). They can be validated if coupled with existing conventional methods (Di Stefano et al. 2005; Porto et al. 2014b; Porto and Walling 2015).

The ^{137}Cs fallout radionuclide (with a half-life of ca. 30 years) is associated with the testing of nuclear weapons occurring from the late 1950s to the early 1960s (Ritchie and Ritchie 2005; Mabit et al. 2014). The released ^{137}Cs was spread to stratosphere, circulated around the globe and during the following years it was gradually deposited on land surface mainly by wet (and partially also dry) fallout. Because of high chemical reactivity of cesium, the deposited ^{137}Cs was quickly bound to soil colloids in the uppermost soil layer and since then it is stable and essentially non-exchangeable (Mabit et al. 2014). It can move only together with soil particles when the soil is moved by physical processes such as soil erosion. Therefore, ^{137}Cs is being used as soil erosion tracer. The methodological guidance of this method is provided by Zapata (2002), IAEA (2014) and Fulajtar et al. (2017).

The principle of this technique is the comparison of ^{137}Cs contamination at a studied site with ^{137}Cs contamination at a so called ‘reference site’ which is not affected by erosion or deposition. The ^{137}Cs inventories obtained from a study area ($^{137}Cs_{Inv}$) are compared with the ^{137}Cs reference value ($^{137}Cs_{Ref}$) obtained from an undisturbed area (reference site) located in the vicinity of the study area. If the values of $^{137}Cs_{Inv}$ are lower than $^{137}Cs_{Ref}$, it means that erosion had occurred, while if the values of $^{137}Cs_{Inv}$ are greater than $^{137}Cs_{Ref}$ it indicates deposition. The magnitude of soil erosion and deposition rate is established by conversion models that are described by a set of different equations depending on the type of soil and land use (Walling and He 1999). For several decades this technique was applied in many areas of the world, and it has several important advantages as compared to conventional measurements:

1. Obtaining retrospective information on long-term mean soil redistribution rates after a single field visit or short (few days to few weeks) sampling campaign. This is probably the most important advantage of FRN techniques as compared

to erosion monitoring programs based on erosion measurements in experimental plots or catchments. These conventional methods need permanent monitoring stations and regular maintenance and operation of measurements usually requiring several staff members and thus are very expensive and labor intensive.
2. Expressing a spatial distribution of the data derived from a grid of sampling points. Erosion plots and catchment studies using hydrological methods refer the erosion rates only to a single point where the measuring device is installed.
3. Estimating the overall soil redistribution rates representing all processes running at the sampled site (water, wind, tillage erosion, etc.).
4. Possibility to estimate the sediment budgets for large areas especially if sedimentation sites like floodplains and reservoirs are surveyed and periodically monitored.

There are, however, some important limitations associated with using ^{137}Cs tracer:

1. The estimates of soil redistribution rates are time related. This means that the derived estimates of soil erosion and deposition are mean values representing the period since the commencement of nuclear fallout (1954) to the time of sampling. This is a very long period making this technique unsuitable to detect changes in erosion rates caused by land use changes and climate changes because during such long period land use and climate usually are not stable.
2. The uncertainty related to the choice of a conversion model able to convert the loss or gain of ^{137}Cs into the estimates of soil loss or deposition can arise due to the lack of input data. The conversion models need calibration and validation using local parameters and these parameters are difficult to express, especially if they should represent long period. This limitation is often overlooked or ignored in many studies.
3. Great spatial variability of ^{137}Cs distribution and the difficulties with representativeness of reference sites.

After several decades of using the ^{137}Cs technique some authors realized that it would be very useful to resample some sites which were studied already earlier, and to compare the rates of soil erosion representing different periods. This allowed them to compare whether the erosion rates increased or decreased since the first sampling. Such approach has been used since the '80s by several authors.

When more and more time has elapsed since the ^{137}Cs fallout in early 1960s the time period represented by ^{137}Cs technique was growing and it became too long for the needs of erosion investigations aimed on developing soil conservation and sustainable land management. While at the beginning of 1990s, when the ^{137}Cs technique became well established and routinely used method, the obtained mean erosion rates represented the period of approximately 30 years, now they represent approximately 60 years. During such long time the land use and land management are developing and may undergo significant changes, so the interpretation of the measured results in relation to land use and land management impacts become difficult. This is one of the main reasons, why recently there is a great need to use the ^{137}Cs resampling approach.

Introducing the ^{137}Cs resampling method is a great achievement and it can be considered as a milestone in erosion research because this approach upgrades the ^{137}Cs method to much higher level and allows the full exploitation of the information potential of radionuclide tracers for soil erosion. It avoids significant part of methodological limitations of ^{137}Cs technique, especially reducing the uncertainties associated with usual applications of ^{137}Cs technique based on one sampling and comparing studied site with reference site. But the major advantage of the resampling method is that it allows controlling the length of the period to which the resulting mean soil erosion rates are associated. For the single sampling approach the investigated period is different for each study (since ^{137}Cs fallout until the time of sampling depending on when it was done), and it is still growing. The resampling approach allows to control the number and length of time periods called here the *time windows*. The period since the ^{137}Cs fallout until presence can be divided by two or more sampling campaigns to two or more time windows what allows to study the temporal dynamics of soil erosion (by comparing the erosion rates of two different time windows). In principle three time windows can be distinguished:

- Whole study period (since the ^{137}Cs fallout until the second sampling)
- First time window (since the ^{137}Cs fallout until the first sampling)
- Second time window (since the first sampling until the second sampling).

Although the first ^{137}Cs resampling studies were done already several decades ago, this approach is still used very rarely and up to now this approach was used only in few tens of study areas due to several reasons, as given below:

- It is not usual to work with earlier (sometimes much older) sample sets.
- It is not obvious to have access to complete primary data and all background data of earlier studies.
- It is not easy to find earlier sampled site which would be well documented (especially the positions of sampling point must be properly identified).
- It is not easy to find earlier sampled site, which would be optimal for the objectives of the new study, because studies have variable objectives and hence the experimental designs were different.

These problems can be easily overcome if the resampling study is done by the authors of the original study. In such case the authors know their own work and all information from the site and the first sampling are available for them. However, such cases are rather rare, because of the short-term nature of research funding, great mobility of the staff and quick changes of the research trends.

For all these reasons the resampling method is developing slowly and the potential of FRN techniques is not fully exploited. Therefore, enhancing this method is a challenge, but it is an important task for erosion research community. This book attempts to improve the ^{137}Cs resampling procedures, provide detailed guidance on how to implement this method, and encourage researchers to use it as much as possible instead of the single sampling approach. The recommended ^{137}Cs resampling approach is based on detailed and systematical reviewing of earlier resampling studies available in literature and on running own studies using ^{137}Cs resampling.

The group of researchers preparing this publication comprise of authors of earlier and recently carried ^{137}Cs resampling case studies in various geographical conditions all over the world.

The developing of this guidelines was based on reviewing the existing literature and assessing the methods and approaches of ^{137}Cs resampling used by different authors. The reviewed studies available in literature present various approaches of resampling, what affects the interpretation of obtained data and the accuracy of final soil redistribution estimates.

In total 14 papers were collected from literature and because some of them presented the results from several study sites finally 54 investigated study sites were reviewed. The studies collected from literature were assessed and the gained experience was exploited in 6 own recent case studies carried by the authors of this guidelines. A list of all studies discussed in this guidelines is shown in Table 1.1 providing also an overview of some basic metadata (country, citation, study area, study site, land use, approach used, and sampling time).

All studies collected from the literature as well as own studies (presented in Chapter 2) were sorted according to their geographical provenience and are arranged in four sections (Northern America and Australia; Europe; Africa; Asia). Australia was discussed together with Northern America because of similar cultural roots, technological capabilities and closely related scientific school, research institutional basis and intensively interacting community of researchers of these English-speaking countries (USA, Canada and Australia).

This regional approach was most appropriate as the studies from certain region may reflect similar environmental conditions, cultural traditions and economical background determining the land use and management, research school and the applied methods. This approach was most suitable for easier assessment and classification of the studies. The old studies collected from literature were characterized briefly, while the new own studies are presented in full detail, and finally all studies are assessed together.

The regional overview of ^{137}Cs resampling studies starts with Subchapter 2.1 presenting the studies carried in North America and Australia. They represent the largest and oldest group of studies as the ^{137}Cs resampling was introduced in these countries. It includes the studies of three groups of authors also involving those who introduced the ^{137}Cs resampling method and made some basic premises and assumptions used later by other authors:

1. Contributions from Canada (Kachanoski and de Jong 1984; Kachanoski 1987; de Jong and Kachanoski 1988; Lobb and Kachanoski 1999; Tiessen et al. 2009; Li et al. 2011)
2. Contributions from the USA (McHenry and Bubenzer 1985; Fornes et al. 2005)
3. Contributions from Australia (Loughran and Balog 2006)

Subchapter 2.2 provides contributions from Europe:

1. Contributions from Italy (Porto et al. 2001, 2003, 2004, 2014a, 2018 and new case studies in this book)

Table 1.1 List of the studies reviewed and assessed in this book

Country	Citation	Study area	Site name	Land use	Re-sampling method		Sampling year	
					SRA	TRA	First	Second
Canada	Kachanoski and de Jong (1984)	Saskatchewan	13 sites (Saskatoon Area)	Cultivated	Authors[a]	Data reprocessed[b]	1966	1981
	de Jong and Kachanoski (1988)	Saskatchewan	26 sites (Saskatoon Area)	Cultivated	Authors	Data reprocessed	mid '60s	early '80s
	Tiessen et al. (2009)	New Brunswick	Grand Falls (20NB)	Cultivated	Authors	Data reprocessed	1990	2005
	Lobb and Kachanoski (1999)	Southeastern Ontario	Brant County Field	Hay pasture/ cultivated	Authors	Not possible[c]	1971	1986
	Li et al. (2011)	Southeastern Ontario	Brant County Field	Hay pasture/ cultivated	Authors	Not possible[c]	1971	1986
	Kachanoski (1987)	Southern Ontario	Guelph (10 plots)	Cultivated	Authors	Data reprocessed	1965	1976
USA	McHenry and Bubenzer (1985)	White Clay Lake Watershed (Wisconsin)	Brunner Farm Field (4 sites)	Cultivated	Not possible	Data reprocessed	1974–1976	1980
	Fornes et al. (2005)	Iowa	Watershed W-11	Cultivated	Authors	Authors	1974	1998
Australia	Loughran and Balog (2006)	New South Wales	Maluna Vineyards (Hunter Valley)	Cultivated	Data re-processed	Data reprocessed	1984–1985	2004

(continued)

Table 1.1 (continued)

Country	Citation	Study area	Site name	Land use	Re-sampling method		Sampling year	
					SRA	TRA	First	Second
Europe (Italy)	Porto et al. (2014a)	Crotone (Calabria)	EOW	Eucalyptus plantation	Not applied[d]	Authors	1998–1999	2013
	Porto et al. (2018)	Crotone (Calabria)	RLW	Rangeland	Not applied[d]	Authors	2001	2014
	This book	Crotone (Calabria)	EFW	Eucalyptus plantation	Not applied[d]	Authors	1999–2001	2014–2015
	This book	Bagnara (Calabria)	Erosion plots (Caratozzolo farm)	Cultivated	Authors	Authors	2006	2017
Europe (Spain)	This book	Pre-Pyrenees (northeast Spain)	The Sola	Forest (hilltop) and cultivated (bottom)	Comparison between two-windows traditional approaches[e]		2003	2016
Africa (Morocco)	Moustakim et al. (2019)	Tangier-Tetouan Region (Northern Morocco)	Nakhla watershed (NA1)	Cultivated	Comparison between two-windows traditional approaches[e]		2002	2018
	Benmansour et al. (2019)	Tangier-Tetouan Region (Northern Morocco)	El Hachef watershed (HA1, HA2)	Cultivated	Comparison between two-windows traditional approaches[e]		2001	2017
	Benmansour et al. (2019)	Tangier-Tetouan Region (Northern Morocco)	Raouz watershed (RA1, RA2)	Cultivated	Comparison between two-windows traditional approaches[e]		2001	2018

(continued)

Table 1.1 (continued)

Country	Citation	Study area	Site name	Land use	Re-sampling method		Sampling year	
					SRA	TRA	First	Second
	This book	Rabat-Kénitra (Northern Morocco)	Rabat-Kénitra (MA1)	Oak plantation	Comparison between two-windows traditional approaches[e]		2007	2019
Asia (China)	Chappel et al. (2015)	Chinese Loess Plateau (Shaanxi province)	Yangjuangou Reservoir catchment	Cultivated (converted to forest/shrubs)	Authors	Not applied[e]	1996	2012
	This book	Chinese Loess Plateau (Shaanxi province)	Yangjuangou Reservoir catchment	Cultivated	Authors	Not applied[e]	1996	2019
	This book	Chinese Loess Plateau (Shaanxi province)	Yangjuangou Reservoir catchment	Forest plantation	Authors	Not applied[e]	1996	2016

[a]The resampling method was applied directly by the original authors; [b]the application of the resampling method required a further assessment; [c]the information reported in the original work did not allow the application of that resampling method; [d]the authors did not apply that resampling method because it was not considered reliable in that specific case; [e]a further step is necessary to apply the resampling approach

2. Contribution from Spain (A. Navas team's new case study in this book)

The first section of Subchapter 2.2 presents three contributions from Italy carried out by the team of Assoc. prof. P. Porto from University Mediterranea of Reggio Calabria (Porto et al. 2001, 2003, 2004). More specifically, the first and the second example include two experimental catchments (EOW and RLW, Crotone Experimental Area) located in Southern Italy, for which some results have already been published (Porto et al. 2014a, 2018). These results will be discussed in detail considering information provided in Subchapter 2.1. The third example contains a new case study that explores the use of this technique on a third experimental catchment (EFW, Crotone Experimental Area). In this last experiment, some of the assumptions made in the previous contributions will be tested and discussed.

The second section of Subchapter 2.2 provides contribution from Spain. It is a new case study carried out at *The Sola* site, an upland agroecosystem located in the central part of the Pre-Pyrenees (Northeast Spain) carried by the team of Dr. L. Gaspar and Dr. A. Navas from Spanish National Research Council. It involves two units with different slope angles, vegetation cover, and soil management. More specifically, the hilltop is under forest (mean slope of 19%) and the lower slope is under arable land where the agricultural fields occupy most of the gentle slope (mean slope of 4.5%).

Subchapter 2.3 provides data from Asia, namely from China, represented by a recent case study carried out by the research team of Dr. H. Q. Yu from Chinese Academy of Agricultural Sciences (CAAS). In this contribution, the Yangjuangou Reservoir Catchment, located in the middle region of the Loess Plateau, was selected as a study site. It includes two parts: the first part presents an experience, made on a cultivated hillslope, in which the ^{137}Cs resampling approach was applied to detect soil erosion change over time in response to land use change; the second part reports two case studies, one from a cultivated land and one from a forest, in which the ^{137}Cs resampling approach was applied to detect changes in soil erosion rates due to climate change.

Finally, Subchapter 2.4 presents the work carried out in Africa represented by studies done in Morocco by the research team of Dr. M. Benmansour from Centre National de l'Energie, des Sciences et des Techniques Nucléaires (CNESTEN). The study areas are in the Tangier-Tetouan Region and the Rabat-Kénitra Region (Northern Morocco). Overall, six study sites were involved. Five of them are cultivated (HA1, HA2, NA1, RA1, RA2) and belong to three watersheds El Hachef, Nakhla and Raouz located in the Tangier-Tetouan Region, and one is forested and belongs to Mâamora Forest in the Rabat-Kénitra Region.

The assessment of collected studies showed up that the resampling method can be based on different assumptions and can use different approaches. Comparing the approaches used in all evaluated case studies showed up that all used approaches can be grouped to two groups representing two major concepts of resampling approach using different reference value for the second sampling campaign:

Traditional reference approach (TRA) uses for the second sampling campaign the reference value from reference site similarly as for the first sampling campaign.

This is the same approach as in the single sampling studies (hence the term 'traditional reference approach'). The reference values for the second sampling campaign can be obtained either by second sampling of reference site, or by just recalculating the values from first reference site sampling (correcting them for radioactive decay). In those areas, where the ^{137}Cs fallout measurements are available the original fallout values can be used as reference value.

Spatial reference approach (SRA) uses as the reference values for second sampling the ^{137}Cs inventories measured at study site during the first sampling. In such case each sampling point has its own point reference value instead of one mean value of reference site. Thus, the reference value for the second period is spatially differentiated (hence the term 'spatial reference approach').

More detailed information on the TRA and SRA is available in Chapter 4.

Chapter 2 providing the regional overview of the studies collected from literature and own case studies is followed by Chapter 3, presenting a recent study from Italy, which is presented as a separate chapter because it is a unique example of the ^{137}Cs resampling technique validation by comparing the ^{137}Cs-based erosion rates with conventional erosion measurements on experimental plots. The erosion rate measurements at experimental plots were used for a validation exercise.

Chapter 4 summarizes the experience from reviewing earlier studies and from own case studies. Key features related to the approaches presented in the previous chapters were taken into account to provide a comprehensive guideline for implementing the ^{137}Cs resampling technique. The 7-step resampling scheme was proposed. It considers available study design fundamentals (background information provided by primary study), possible approaches, and variable objectives of the resampling study and presents two basic data interpretation approaches introduced above (TRA and SRA). Based on that a step-by-step protocol was proposed. This is a major result of this guidelines. It should serve as a ready to use tool for erosion researchers to choose optimal approach for successful implementation of ^{137}Cs resampling technique.

References

Benmansour, M, Moussadek, M, Moustakim, M, Zouagui, A, Nouira, A, Damnati, B, Iaaich, H, Mrabet, R, Fulajtar, E, Mabit, L (2019) Use of fallout radionuclides to assess change in soil erosion and sedimentation rates in Northern Morocco (2019). Food and Organisation of the United Nation. Proceeding of Global Symposium on Soil Erosion, 656–660

Chappell, A, Li, Y, Yu, HQ, Zhang, YZ, Li, XY (2015) Cost-effective sampling of $^{(137)}Cs$-derived net soil redistribution: part 2 - estimating the spatial mean change over time. J Environ Radioact, 144: 168–74

De Jong E, Kachanoski RG (1988) The importance of erosion in the carbon balance of prairie soils. Can J Soil Sci 68:111–119

Di Stefano C, Ferro V, Porto P, Rizzo S (2005) Testing a spatially distributed sediment delivery model (SEDD) in a forested basin by caesium-137 technique. J Soil Water Conserv 60(3):148–157

Fornes WL, Whiting PJ, Wilson CG, Matisoff G (2005) Caesium-137-derived erosion rates in an agricultural setting: the effects of model assumptions and management practices. Earth Surf Process Land 30:1181–1189

Fulajtar E, Mabit L, Renschler CS, Lee Zhi Yi A (2017) Use of ^{137}Cs for soil erosion assessment. FAO, 542 Rome, 76 pp

International Atomic Energy Agency (2014) Guidelines for using fallout radionuclides to assess erosion and effectiveness of soil conservation strategies. IAEA-TECDOC-1741. IAEA, Vienna, 224 pp

Kachanoski RG (1987) Comparison of measured soil 137-cesium losses and erosion rates. Can J Soil Sci 67:199–203

Kachanoski RG, de Jong E (1984) Predicting the temporal relationship between cesium-137 and erosion rate. J Environ Qual 13:301–304

Li S, Lobb DA, Kachanoski RG, McConkey BG (2011) Comparing the use of the traditional and repeated-sampling-approach of the Cs-137 technique in soil erosion estimation. Geoderma 160(3–4):324–335

Lobb DA, Kachanoski RG (1999) Modelling tillage erosion in the topographically complex landscapes of southwestern Ontario, Canada. Soil Tillage Res 51:261–277

Loughran RJ, Balog RM (2006) Re-sampling for soil-caesium-137 to assess soil losses after a 19-year interval in a Hunter Valley vineyard, New South Wales, Australia. Geogr Res 44(1):77–86

McHenry JR, Bubenzer GD (1985) Field erosion estimated from ^{137}Cs activity measurements. Trans Am Soc Agric Eng 28:480–483

Mabit L, Chhem-Kieth S, Dornhofer P, Toloza A, Benmansour M, Bernard C, Fulajtar E, Walling DE (2014) ^{137}Cs: a widely used and validated medium term soil tracer. In: IAEA. Guidelines for using fallout radionuclides to assess erosion and effectiveness of soil conservation strategies, IAEA-TECDOC-1741, 27–77

Mabit L, Bernard C, Lee A, Yi Z, Fulajtar E, Dercon G, Zaman M, Toloza A, Heng L (2018) Promoting the use of isotopic techniques to combat soil erosion: an overview of the key role played by the SWMCN Subprogramme of the Joint FAO/IAEA Division over the last 20 years. Land Degrad Dev 29(9):3077–3091

Moustakim, M, Benmansour, M, Zouagui, A, Nouira, A, Benkdad, A, Damnati, B (2019) Use of caesium-137 re-sampling and excess lead-210 techniques to assess changes in soil redistribution rates within an agricultural field in Nakhla watershed. Journal of African Earth Sciences, 156, 158e167

Panagos P, Standardi G, Borrelli P, Lugato E, Montanarella L, Bosello F (2018) Cost of agricultural productivity loss due to soil erosion in the European Union: from direct cost evaluation approaches to the use of macroeconomic models. Land Degrad Dev. 29:471–484

Pimentel D, Harvey C, Resosudarmo P, Sinclair K, Kurz D, McNair M, Crist S, Shpritz L, Fitton L, Saffouri R, Blair R (1995) Environmental and economic costs of soil erosion and conservation benefits. Science, New Series 267(5201):1117–1123

Porto P, Walling DE (2015) Use of caesium-137 measurements and long-term records of sediment load to calibrate the sediment delivery component of the SEDD model and explore scale effect: examples from Southern Italy. J Hydrol Eng. 20(6):C4014005

Porto P, Walling DE, Alewell C, Callegari G, Mabit L, Mallimo N, Meusburger K, Zehringer M (2014a) Use of a ^{137}Cs re-sampling technique to investigate temporal changes in soil erosion and sediment mobilisation for a small, forested catchment in southern Italy. J Environ Radioact 138:137–148

Porto, P, Walling, DE, Callegari, G (2004) Validating the use of caesium-137 measurements to estimate erosion rates in three small catchments in Southern Italy. IAHS Publication, 288:75–83

Porto P, Walling DE, Callegari G (2018) Using repeated ^{137}Cs and $^{210}Pb_{ex}$ measurements to establish sediment budgets for different time windows and explore the effect of connectivity on soil erosion rates in a small experimental catchment in Southern Italy. Land Degrad Dev 29:1819–1832

Porto P, Walling DE, Capra A (2014b) Using ^{137}Cs and $^{210}Pb_{ex}$ measurements and conventional surveys to investigate the relative contributions of interrill/rill and gully erosion to soil loss from a small, cultivated catchment in Sicily. Soil Tillage Res 135:18–27

Porto, P, Walling, DE, Ferro, V (2001) Validating the use of caesium-137 measurements to estimate soil erosion rates in a small drainage basin in Calabria Southern Italy Journal of Hydrology 248(1–4):93–108

Porto, P, Walling, DE, Ferro, V, Di Stefano, C (2003) Validating erosion rate estimates by caesium-137 measurements for two small forested catchments in Calabria, southern Italy. Land Degrad. Develop. 14:389–408

Ritchie JC, Ritchie CA (2005) Bibliography of publications of 137-cesium studies related to erosion and sediment deposition. USDA-ARS Hydrology and Remote Sensing Laboratory. Occasional Paper HRSL-2005-01

Tiessen KHD, Li S, Lobb DA, Mehuys GR, Rees HW, Chow TL (2009) Using repeated measurements of ^{137}Cs and modelling to identify spatial patterns of tillage and water erosion within potato production in Atlantic Canada. Geoderma 153(1–2):104–118

Walling DE, He Q (1999) Improved models for estimating soil erosion rates from cesium-137 measurements. J Environ Qual 28:611–622

Zapata F (ed) (2002) Handbook for the assessment of soil erosion and sedimentation using environmental radionuclides. Kluwer, Dordrecht, 219 pp

Chapter 2
Geographical Overview of ^{137}Cs Resampling Studies

P. Porto, E. Fulajtar, D. E. Walling, G. Callegari, V. Cogliandro, C. La Spada, L. Gaspar, A. Navas, H. Q. Yu, W. X. Liu, A. Chappell, Y. Li, M. Benmansour, M. Moustakim, B. Damnati, R. Moussadek, A. Nouira, N. Amenzou, R. Mrabet, and L. Kheng Heng

Abstract In this chapter, all studies on ^{137}Cs resampling collected from the literature as well as the recent contributions carried out by the authors of this guidelines are presented. The grouping of assembled studies was based on their geographical provenience. All discussed contributions were arranged into four subchapters: 2.1 North America and Australia (English speaking countries Canada, USA, and Australia), 2.2 Europe (represented by Italy and Spain), 2.3 Asia (represented by China) and 2.4 Africa (represented by Morocco). Each subchapter is further subdivided to countries and each country to individual studies (or groups of studies carried by one research team). The main purpose is to discuss and evaluate the reviewed studies and to classify the used ^{137}Cs resampling approaches.

In this chapter, all studies on ^{137}Cs resampling collected from the literature as well as the recent contributions carried out by the authors of this guidelines are presented. The assessed studies offer a wide spectrum of sampling methods, sampled periods, spatial scales, land uses, soil types, conversion models, magnitude of erosion rates etc. and their systematical grouping is difficult. For this reason, it was agreed that grouping the contributions according to their geographical provenience was the most appropriate approach to classify them and to discuss the pros and cons separately.

P. Porto (✉) · V. Cogliandro · C. La Spada
Dipartimento Di Agraria, University Mediterranea of Reggio Calabria, Reggio Calabria 89122, Italy
e-mail: paolo.porto@unirc.it

P. Porto · G. Callegari
CNR. Istituto per i Sistemi Agrari e Forestali per il Mediterraneo, Sezione Ecologia e Idrologia Forestale, Rende, Italy

E. Fulajtar · L. Kheng Heng
Joint FAO/IAEA Centre of Nuclear Techniques in Food and Agriculture, International Atomic Energy Agency, Vienna, Austria
e-mail: emil.fulajtar@nppc.sk

P. Porto et al. (eds.), *Using ^{137}Cs Resampling Method to Estimate Mean Soil Erosion Rates for Selected Time Windows*, Environmental Science and Engineering,
https://doi.org/10.1007/978-3-031-52807-1_2

Following this logic, the contributions were arranged into four subchapters: 2.1 North America and Australia (English speaking countries Canada, USA, and Australia), 2.2 Europe (Italy and Spain), 2.3 Asia (China), and 2.4 Africa (Morocco). Each subchapter is further subdivided to countries and each country to individual studies (or groups of studies carried by one research team). In some cases, the basic characterization of the case studies is followed by discussion. This approach was chosen because some of the original publications did not present the estimates of soil redistribution rates for all three possible periods (the whole study period, the first and second time windows), and this should be considered when comparing them with similar works carried out in other study areas. However, in most cases, the potential to do this calculation was exploited and the possibility to derive additional information from other contributions made in the same study areas was considered. This effort was aimed on establishing local reference inventories (where absent), selecting a proper conversion model (where not reported by the authors), and applying both

L. Kheng Heng
e-mail: l.heng@iaea.org

D. E. Walling
Geography, College of Life and Environmental Sciences, University of Exeter, Amory Building, Rennes Drive, Exeter EX4 4RJ, UK

L. Gaspar · A. Navas
Soil and Water Department, Aula Dei Experimental Station, EEAD-CSIC, Avenida Montañana, 1005, Zaragoza 50059, Spain

H. Q. Yu · W. X. Liu · Y. Li
Institute of Environment and Sustainable Development in Agriculture, Chinese Academy of Agricultural Sciences (CAAS), Haidian District, Beijing 100081, China

A. Chappell
School of Earth and Environmental Sciences, Cardiff University, Cardiff, Wales, UK

M. Benmansour · M. Moustakim · A. Nouira · N. Amenzou
Centre National de L'Energie, des Sciences et des Techniques Nucléaires, BP 1382, R.P.10001 Rabat, Morocco

B. Damnati
Faculté des Sciences et Techniques, Département des Sciences de la Terre, BP 416, Tangier, Morocco

R. Moussadek
International Center for Agricultural Research in Dry Areas, Av. Hafiane Cherkaoui, Rabat, Morocco

R. Moussadek · R. Mrabet
Institut National de la Recherche Agronomique, 415, Rabat, Morocco

E. Fulajtar
Soil Science and Conservation Research Institute, National Agricultural and Food Centre, Bratislava, Slovakia

P. Porto
Faculty of Geographical Sciences, Kazimierz Wielki University, Bydgoszcz, Poland

the traditional reference approach (TRA) and the spatial reference approach (SRA) (where only one approach was used). This suplementary work required additional processing of the original datasets to harmonize the results and to make them useful for comparison with other case studies. Further, it should be pointed out, that these case studies use different equations and conversion models to calculate soil redistribution rates, and variable terminology to denominate parameters used in equation as well as the time windows to which the calculated erosion rates are attributed. This terminology is preserved in its original form for all case studies in Chapter 2 in order to keep these studies authentic. Therefore, it should be kept in mind that terminology was not standardised as in Chapters 1, 3 and 4 and in different case studies the same parameter can have different designation, or the same term can have different meaning. For example, in some studies the term "second time window" is not used for time window since the first until the second sampling but since the ^{137}Cs fallout until the second sampling, etc. Nevertheless, all terms are defined and explained in each case study where they are used so the different meaning should be well understood. At the end of each subchapter, a discussion is provided to outline the advantages of the used approaches and to address the eventual weak points.

2.1 Using ^{137}Cs Resampling Method in North America and Australia: A Review

In this chapter, the existing studies carried out in Canada, USA, and Australia are reviewed. The information provided in the original works is reviewed and the main key features and assumptions made by the authors are discussed and further interpreted. Some of these results will be recalled in Chapter 4 in which a step-by-step protocol for the general application of the technique is proposed.

2.1.1 Contributions from Canada

In Canada, the pioneering studies relating soil erosion and ^{137}Cs were conducted during the early 1980s by de Jong et al. (1982, 1983). The technique, applied in its traditional form, was successfully used across the country by many other authors during the following years (see among the others, Bernard and Laverdiere 1992; Cao et al. 1994; Zebarth et al. 2002). Several years after the first applications, some of these authors pioneered the resampling approach revisiting some of the sites that were previously sampled for ^{137}Cs measurements (Fig. 2.1).

The examples of resampling approach are reported by Kachanoski and de Jong (1984), Kachanoski (1987), de Jong and Kachanoski (1988), Lobb and Kachanoski (1999), Tiessen et al. (2009), and Li et al. (2011).

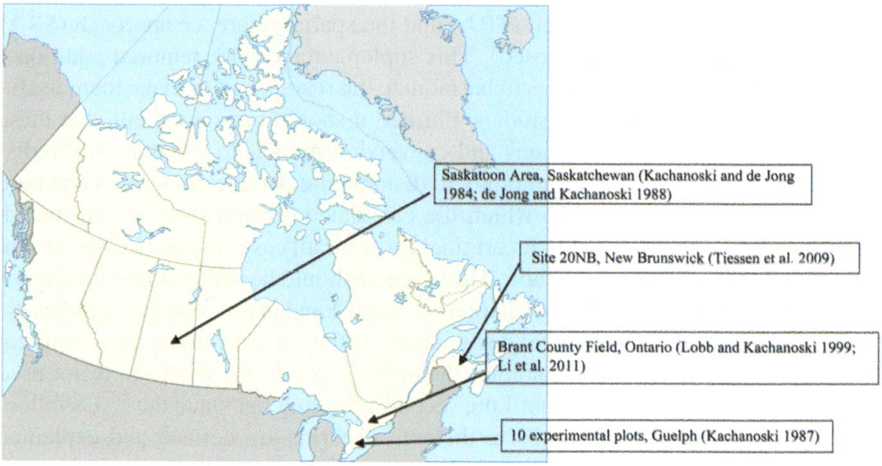

Fig. 2.1 Experimental sites in CANADA that have been resampled for ^{137}Cs analysis by Kachanoski and de Jong (1984), Kachanoski (1987), de Jong and Kachanoski (1988), Lobb and Kachanoski (1999), Tiessen et al. (2009), and Li et al. (2011)

2.1.1.1 Kachanoski and De Jong (1984) Approach

The resampling approach reported by Kachanoski and de Jong (1984) was probably the first attempt to document the advantage of repeating soil sampling for ^{137}Cs measurements to estimate soil erosion rates. Their work is based on two sets of measurements made in the Saskatoon Area over 13 cultivated sites. The sites were firstly sampled in 1966 and then resampled in 1981 (after 15 years). For each site and for each sampling campaign, five soil cores (7.6 cm diameter) were collected and bulked to get well representative composite sample for ^{137}Cs analyses. The ^{137}Cs activity was determined using the laboratory procedure described by de Jong et al. (1982). The samples collected in 1966 were analyzed in 1981 and corrected for decay. A ^{137}Cs reference value (2368 Bq m^{-2}) was measured in 1980 by de Jong et al. (1982). This value was compared with that provided, for the same year, by the fallout measurements available in Saskatchewan (see Kachanoski and de Jong 1984). The latter, corrected for local conditions, were used as input for the following model:

$$\frac{\partial A_t}{\partial t} = D_t - E_t k_2 \left(\frac{A_t}{M} \right) - k_1 A_t \tag{2.1}$$

where:

t is time (yr);
A_t is the total ^{137}Cs inventory existing in plough layer (Bq m^{-2});
E_t is the average rate of soil erosion (kg m^{-2} yr^{-1});
k_1 is the radioactive decay for ^{137}Cs (0,023 yr^{-1});

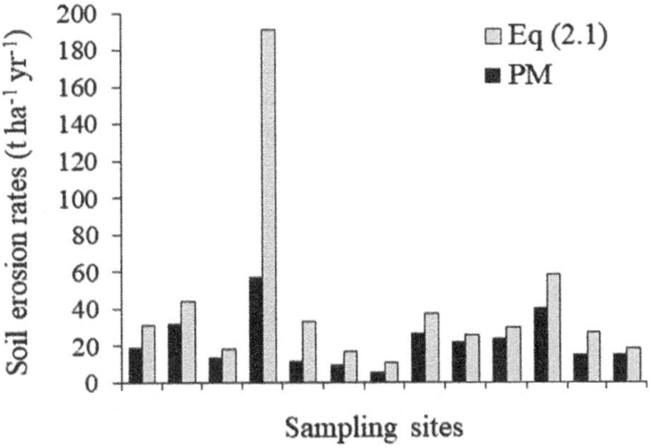

Fig. 2.2 Comparison between the soil loss for the 1967–1981 period obtained with the model proposed by Kachanoski and de Jong (1984) and the Proportional Model. (data from Kachanoski and de Jong, 1984)

k_2 is an enrichment coefficient (here assumed to 1.1);
M is the mass of soil in plough layer (here assumed to 100 kg m^{-2});
D_t is the fallout input (Bq m^{-2}).

The Eq. 2.1. was integrated over time to get the estimates of soil erosion and deposition over the ith time period (see for details Kachanoski and de Jong 1984). The estimates of soil erosion (kg m^{-2} yr^{-1}) obtained with their proposed model, are derived using the spatial reference approach (SRA) defined in Chapter 1. The erosion rates are reported in their original work (see Table 2 in Kachanoski and de Jong 1984) and compared with the corresponding estimates obtained with the proportional model (PM).

A comparison between the estimates of soil erosion calculated by two models (Fig. 2.2), suggests that the model expressed by the Eq. 2.1, proposed by the authors, provides higher estimates of soil erosion (mean of 41.6 t ha^{-1} yr^{-1}) than those obtained with the PM (mean of 22.5 t ha^{-1} yr^{-1}) for all 13 field sites. Also, Fig. 2.2 indicates that the application of SRA shows the absence of deposition in the study area.

Discussion: A key feature of the basic approach suggested by Kachanoski and de Jong is related to the fact that the period between the two sampling campaigns (ca 15 years) could be long enough to provide detectable differences between the corresponding ^{137}Cs inventories. In their original work, the authors reported the ^{137}Cs measurements, for each site and for each sampling campaign, in terms of ^{137}Cs concentration of the plough layer. Using these data and a mass depth equal to 100 kg m^{-2}, it is possible to obtain the total inventory for each site. If we plot these inventories against the 1:1 line for comparison (Fig. 2.3), it is clear that the 1966 inventories (decay-corrected to 1981) are all higher than the corresponding values related to the sampling undertaken in 1981. This sensitive difference suggests that the time span occurring between the two sampling campaigns may be long enough to provide suitable estimates of soil erosion.

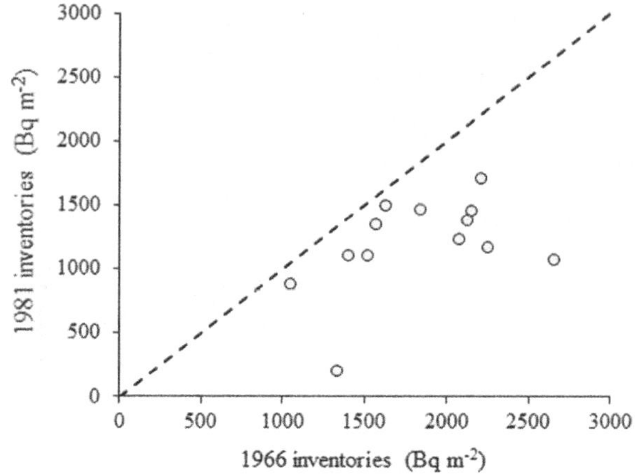

Fig. 2.3 Comparison of the corresponding ^{137}Cs inventories for the same sampling sites related to the sampling campaigns in 1966 and 1981. The 1966 inventories are decay corrected to 1981

The data reported by the authors can also be combined with the measurements of fallout to get the reference value of ^{137}Cs for the two sampling campaigns. Based on the assumption that in the Saskatoon Area the atmospheric ^{137}Cs deposition was calculated as 84% of those provided for Saskatchewan, the resulting reference values are ca 2520 (Bq m^{-2}) for 1966 and ca 2327 (Bq m^{-2}) for 1981. Using a simple PM, these estimates can be used for the application of the traditional reference approach (TRA) to obtain a long-term estimate of erosion rates for the 1954–1966 and 1954–1981 periods respectively.

For the calculation, a plough layer of 0.1 m, a bulk density of 1000 kg m^{-3}, and time equal to 12 and 27 years, for 1966 and 1981 respectively, were used. The resulting estimates of soil redistribution rates are reported in Fig. 2.4.

The

obtained estimates (Fig. 2.4) confirm the absence of deposition for the period comprised between the two sampling campaigns. However, the mean erosion rate

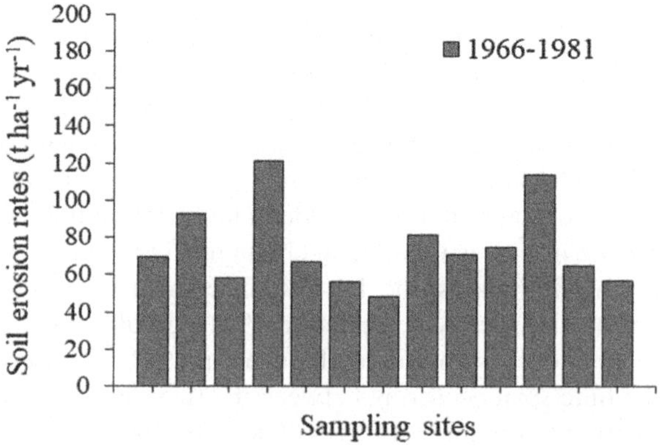

Fig. 2.4 Estimates of soil redistribution rate for the 1967–1981 period using the TRA

(75 t ha^{-1} yr^{-1}) obtained coupling the TRA with the PM was higher than that obtained using the SRA coupled with the Eq. 2.1 (41.6 t ha^{-1} yr^{-1}).

In general, it is difficult to establish which approach provides the best estimate of soil erosion for that area because no independent measurements of soil loss are available. However, it is important to recognize that for these study sites the resulting estimates of soil erosion are strongly dependent on the approach employed.

2.1.1.2 Kachanoski (1987) Approach

The study reported by Kachanoski (1987) was probably the first attempt to compare estimates of soil erosion using a resampling approach for ^{137}Cs measurements with direct measurements of erosion rates. The approach is based on two sets of measurements obtained from ten experimental plots (44 m long) established in 1953 in Southern Ontario by the University of Guelph. During a first sampling campaign undertaken in 1965, soil samples for ^{137}Cs analyses were collected from the top and bottom of each plot, involving 20 sampling locations, to a depth of 15 cm. The same 20 points were resampled in 1976 as part of the long-term monitoring of the site. Each sample was a composite of 10 cores (2.5 cm diameter). Additional soil cores (7.6 cm diameter to a depth of 30 cm) were collected in 1985 from adjacent undisturbed sites covered by permanent grassland and forest to establish the ^{137}Cs reference value. In this case, two cores were combined for each of three slope positions in the grassland and forest. A mean value of 2558 Bq m^{-2} was obtained for the grassland while a mean value of 2662 Bq m^{-2} was obtained for the forested area. These two values compare well with an estimate of the reference value (2700 Bq m^{-2}) derived by the author for 1985 using ^{137}Cs fallout measurements in Ontario (see Kachanoski 1987). All the samples collected in 1965 and 1976 were decay corrected to the same year (1985). For each sampling campaign, the same laboratory procedure was employed for ^{137}Cs analyses. To get the estimates of soil erosion for the period of 1966–1976, the SRA based on a simple proportional model was used:

$$E = \frac{M_s(Cs_i - Cs_f)}{(t_i - t_f)Cs_i} \tag{2.2}$$

where:

E	is the predicted erosion rate (kg m^{-2} yr^{-1});
M_s	is specific mass of the plough layer (kg m^{-2}), calculated from the average bulk density (1300 kg m^{-3}) and the average depth of the plough layer (0.15 m);
t_i and t_f	is time of initial and final soil ^{137}Cs sampling, respectively (yr);
Cs_i and Cs_f	is total soil ^{137}Cs corrected for radioactive decay at time t_i and t_f respectively (Bq m^{-2}).

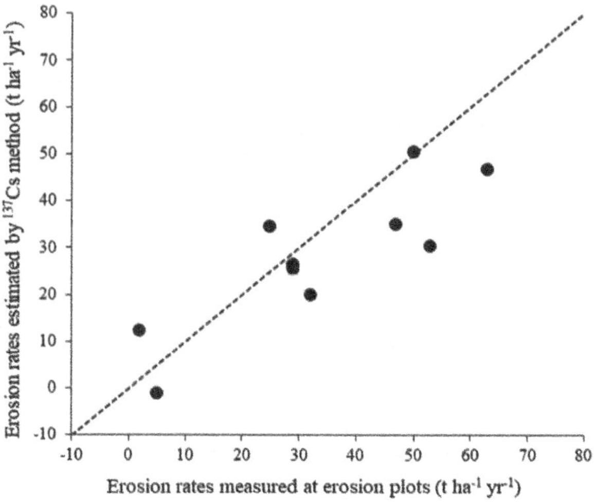

Fig. 2.5 Comparing soil erosion rates measured at erosion plots and estimated by ^{137}Cs method for the 1966–1976 period. (data from Kachanoski 1987, replotted)

The estimates of soil redistribution rates derived from the Eq. 2.2 are compared on Fig. 2.5 with the measurements of soil loss obtained from the ten experimental plots for the 1966–1976 period.

Fig. 2.5 suggests a satisfactory agreement between measured and predicted soil loss using the SRA. The slight underestimation provided by the ^{137}Cs technique can be due to the assumed values of bulk density and plough depth or to the error related in the estimate of reference value.

Discussion: The estimates of soil erosion obtained by Kachanoski (1987) are in good agreement with independent measurements of soil loss available for the ten experimental plots used in his study, and this encourages the use of the SRA to provide estimates of soil redistribution rate. The results obtained in his work made the experiment in some way unique because the data available for the 10 plots monitored by the University of Guelf are probably the only example of long-term measurements of soil loss available in Canada for the period covered by ^{137}Cs measurements (since 1953). Also, the fact that fallout measurements of ^{137}Cs compare well with the reference inventory measured in the two adjacent reference sites gives more strength to the reliability of this approach. The availability of the reference inventory of ^{137}Cs for the time of both sampling campaigns in 1965 and 1976 makes also possible the application of TRA if an estimate of soil erosion rate for the two time windows related to the two sampling campaigns can be obtained. In fact, based on the information provided by the author, the two ^{137}Cs reference values derived from fallout measurements resulted in ca. 3123 (Bq m^{-2}) and ca. 2996 (Bq m^{-2}), respectively for 1965 and 1976. A simple proportional model can then be employed for this comparison using the information provided by the author about the depth of tillage (ca. 0.15 m) and the soil bulk density (ca. 1300 kg m^{-3}). The year of fallout commencement was set to 1954 for simplification of determining the lower limit of each time window. The comparison is shown on Fig. 2.6.

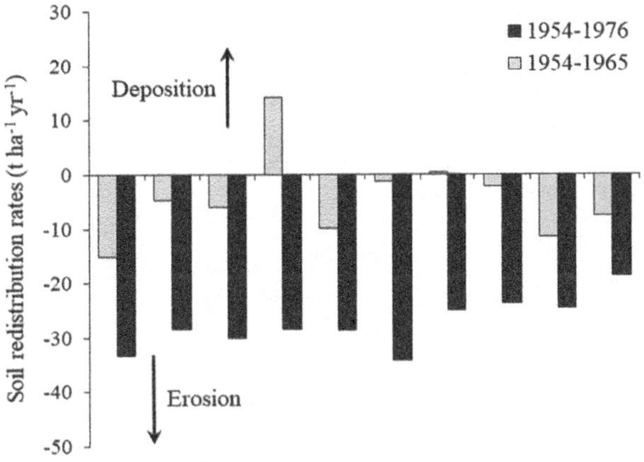

Fig. 2.6 Comparison of the soil redistribution rates for the corresponding sampling sites related to the sampling campaigns in 1960s and 1970s (data from Kachanoski, 1987)

Using the information provided on Fig. 2.6 it is worth noticing that:

1. Two points that showed deposition in 1965 were eroded in 1976.
2. All the measurements made in 1976 were eroded.
3. The rates of net soil erosion obtained in 1965 (mean value of 4.3 t ha^{-1} yr^{-1}) are much lower than those obtained in 1976 (mean value of 27.5 t ha^{-1} yr^{-1}).

This information can be used to apply the TRA by the following steps:

1. The estimates (t ha^{-1} yr^{-1}) obtained for the 1965 sampling campaign were time integrated (t ha^{-1}) for the 1954–1965 period.
2. Accordingly, the estimates (t ha^{-1} yr^{-1}) obtained for the 1976 sampling campaign were time integrated (t ha^{-1}) for the 1954–1976 period.
3. The point-by-point difference between the corresponding estimates obtained above provided the cumulative soil redistribution (t ha^{-1}) for the 1966–1976 period.
4. These values were then divided by the number of years (11) elapsed between the two sampling campaigns and the mean annual soil redistribution rate (t ha^{-1} yr^{-1}) for the 1966-1976 period was obtained.

The final results are compared on Fig. 2.7 with soil erosion rates obtained from the experimental plots for the same period.

It is clear from Fig. 2.7 that the TRA overestimates the erosion rates for the 1966–1976 period. This is in line with the findings derived from the previous paper by Kachanoski and de Jong (1984) that documented higher estimates of erosion rate if the TRA is used.

2.1.1.3 De Jong and Kachanoski (1988) Approach

The resampling approach reported by de Jong and Kachanoski (1988) is based on previous contributions made by the Canadian research group in Saskatchewan (de

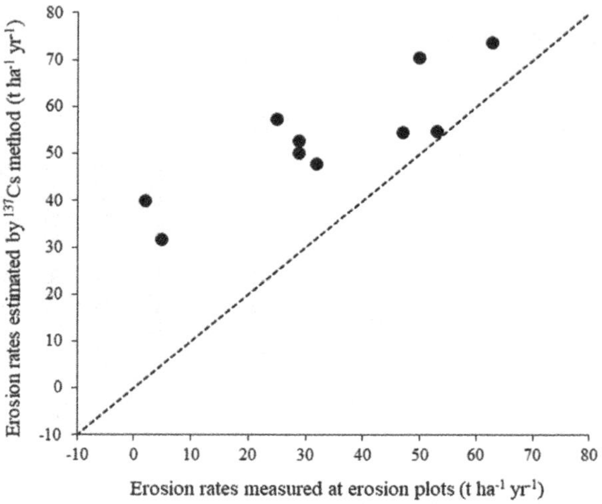

Fig. 2.7 Comparison between erosion rates measured at erosion plots and values estimated by ^{137}Cs method for the 1966–1976 period using the TRA. (data from Kachanoski 1987)

Jong et al. 1982, 1983; Kachanoski and de Jong 1984). This work is built on a first set of soil samples collected in the mid-1960s at 26 cultivated sites. The same sites were resampled in the early 1980s (after ca. 15–20 years) and a new set of samples was available for ^{137}Cs analyses. On both occasions, two bulk soil samples were collected at depths of 0–0.15 m and 0.15–0.30 m at each sampling site. Based on the information reported by the authors (de Jong and Kachanoski 1988), all the samples were analyzed for ^{137}Cs content in the early 1980s, shortly after the second sampling campaign. The ^{137}Cs concentration of the samples collected in the mid-1960s was decay corrected to derive the radionuclide activity at the original time of sampling. In this respect, the authors proposed three ways to calculate the erosion rates for the period between the two sampling campaigns.

The first approach is based on the following mass balance model:

$$\frac{\partial A}{\partial t} = D_t - E_t k_2 C_t + k_1 A_t \tag{2.3}$$

where:

t is time (yr);
A_t is the total ^{137}Cs areal activity density existing in plough layer (Bq m^{-2});
E_t is the average rate of soil erosion (kg m^{-2} yr^{-1});
k_1 is the radioactive decay for ^{137}Cs (yr^{-1});
k_2 is an enrichment coefficient;
C_t is the ^{137}Cs activity in plough layer (Bq kg^{-1});
D_t is the fallout input (Bq m^{-2}).

This approach requires apart of the measurements of ^{137}Cs areal activity for each sampling site, the ^{137}Cs fallout and an estimate of soil erosion rate. The former was calculated using the annual ^{137}Cs deposition measured at Pinawa (Manitoba), provided by Atomic Energy of Canada, while the latter was derived from previous field experience (see de Jong et al. 1982). The Eq. 2.3 is integrated over a time period t using the form proposed by Kachanoski and de Jong (1984).

The second approach is based on the following equation:

$$E_2 = \frac{\left({}^{137}Cs_{1960} - {}^{137}Cs_{1980}\right) \cdot 100}{N \cdot {}^{137}Cs_{1960}} \tag{2.4}$$

where:

E_2 is the rate of soil erosion (expressed as percent of topsoil loss per year);
N is time elapsed between the two sampling campaigns (yr).

The two sets of estimates obtained from Eq. 2.3 and Eq. 2.4, as mentioned by the authors, provided highly correlated estimates of soil erosion, even if a few points divert from the line of perfect agreement.

The third approach is based on the following equation, proposed by de Jong et al. (1983):

$$E_3 = \frac{\left(0.95 \cdot {}^{137}Cs_{\mathrm{REF}} - {}^{137}Cs_{i,1980}\right) \cdot 100}{25 \cdot 0.95 \cdot {}^{137}Cs_{\mathrm{REF}}} \tag{2.5}$$

where:

E_3 is the rate of soil erosion (expressed as percent of topsoil loss per year);
Cs_{REF} is the ^{137}Cs inventory obtained from nearby native grassland sites assumed as a control (Bq m^{-2});
$Cs_{i,1980}$ is the ^{137}Cs inventory of the single points sampled in 1980s during the second sampling campaign in 1980s (Bq m^{-2});
25 is time elapsed since 1960 until the second sampling campaign (yr);
0.95 is a coefficient used by the authors to account for a ^{137}Cs loss due to snow blowing and crop removal.

The estimates obtained with Eq. 2.5 are compared in Fig. 2.8 with the corresponding estimates provided by Eq. 2.4.

It is clear from Fig. 2.8 that the two sets of estimates are poorly correlated. The authors suggested that their third approach is less suitable than the other two because it is based on erosion estimates since 1960 and it does not reflect the time window between the two sampling campaigns. In fact, the inventories obtained for each single point during the second campaign (1980s) were compared with an independent 'reference value' derived from an undisturbed area located in the vicinity of the study site. The resulting estimates are related to a time window (1960–1985) that is longer than time window between the two sampling campaigns and, for this reason, the comparison is not consistent with the other two methods.

Discussion: A key feature of the basic approach suggested by de Jong and Kachanoski is related to the fact that the timespan occurring between the two sets

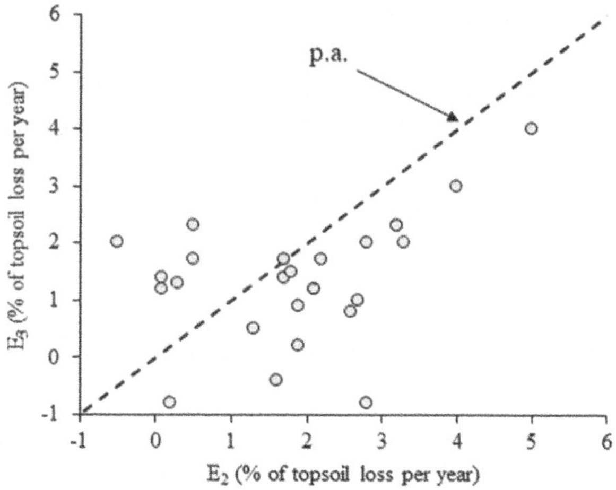

Fig. 2.8 Comparison between the estimates of soil loss using Eq. 2.4 and Eq. 2.5 (data from de Jong and Kachanoski 1988)

of measurements (ca. 20 years) could be long enough to provide appreciable differences between the corresponding ^{137}Cs inventories if the additional fallout occurring during the corresponding period is considered.

The approaches proposed by the authors using Eq. 2.3 and Eq. 2.4 are based on the main assumption that the first set of measurements served as 'reference values' for the second dataset. In other words, the SRA was used to derive the erosion rates for the non-overlapping period.

In their original contribution, the authors did not provide information on ^{137}Cs inventories in uneroded sites close to the study area, even if Eq. 2.5 is based on their values. However, they report the annual ^{137}Cs fallout for the Pinewa station in Manitoba from 1964 to 1985. Combining these data with those provided by Kachanoski and de Jong (1984) for Saskatchewan, it is possible to derive a ^{137}Cs reference inventory for 1965 (3019 Bq m^{-2}) and for 1985 (2597 Bq m^{-2}). These two reference values make the application of the ^{137}Cs technique in its traditional way possible, in order to get mean estimates of soil erosion and deposition for the two-time windows 1954–1965 and 1954–1985. A simple proportional model can be employed for this comparison using the information provided by the authors about the depth of tillage (ca. 0.20 m) and the soil bulk density (ca. 1400 kg m^{-3}), assuming, as mentioned by the authors, that the latter did not change since 1960 until 1980s. The comparison is presented on Fig. 2.9.

Using the information provided in Fig. 2.9 it is worth noticing that:

1. Some of the points that showed deposition in 1960s were eroded in 1980s.
2. Some of the points that showed high values of soil erosion in 1960s had lower rates of erosion in 1980s and vice-versa.

These results suggest that, during the period between the two sampling campaigns, the movement of soil within the sampling sites was considerable. This is better expressed by Fig. 2.10 that is related to the 1966–1985 period only.

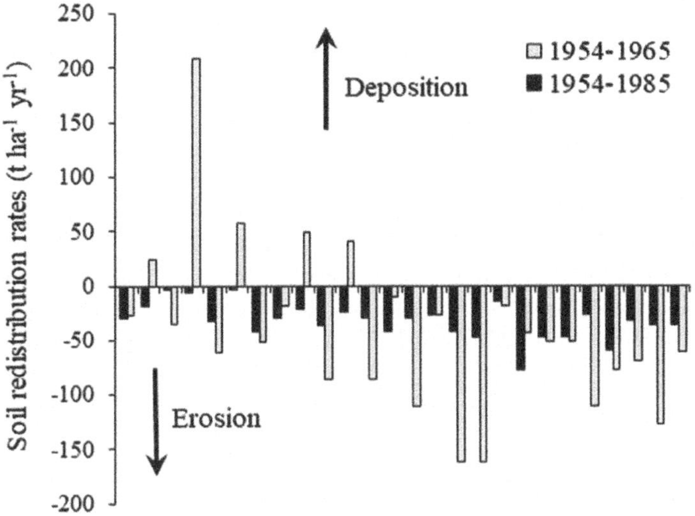

Fig. 2.9 Comparison of the soil redistribution rates for the corresponding sampling sites related to the sampling campaigns during the 1960s and the 1980s (data from de Jong and Kachanoski, 1988)

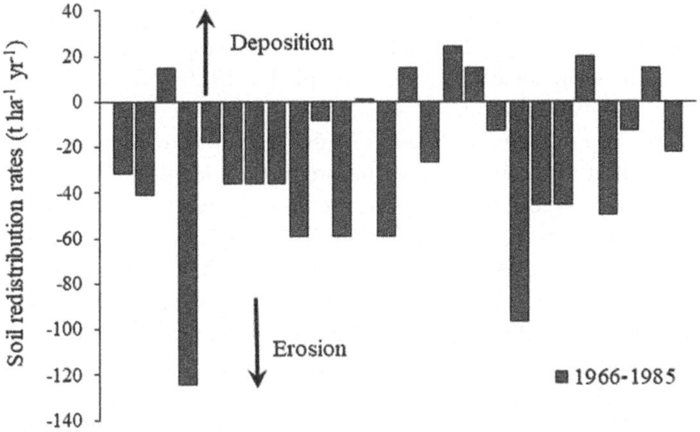

Fig. 2.10 Soil redistribution rates for the 26 sampling sites derived for the 1966–1985 period using the TRA

The information reported on Fig. 2.10 was obtained using the TRA by the following steps:

1. The estimates (t ha^{-1} yr^{-1}) obtained for the 1965 sampling campaign were time integrated (t ha^{-1}) for the 1954–1965 period, assuming that the commencement of ^{137}Cs fallout was in 1954.

2. Accordingly, the estimates (t ha^{-1} yr^{-1}) obtained for the 1985 sampling campaign were time integrated (t ha^{-1}) for the 1954–1985 period.

3. The point-by-point difference between the corresponding estimates obtained above provided the cumulative soil redistribution (t ha^{-1}) for the 1966–1985 period.
4. These values were then divided by the number of years (20) elapsed between the two sampling campaigns and the final estimates of soil redistribution rates (t ha^{-1} yr^{-1}) were obtained.

The data on Fig. 2.10 show that some of the sampling points (7 out of 26) eroded during the 1954–1965 period, showed deposition for the 1966–1985 period. The reversal occurred for the 5 points that showed deposition for the 1954–1965 period (see Fig. 2.10). For the remaining sampling sites, soil erosion is confirmed and showed to be the dominant process in the area. The procedure suggests, for the 1966–1985 period, a net soil loss of ca 27.5 (t ha^{-1} yr^{-1}), which is in rough agreement with the estimate (ca 33 t ha^{-1} yr^{-1}) provided by the authors using Eq. 2.3. These results suggest that both the SRA and TRA lead to similar results for this study site.

2.1.1.4 Lobb and Kachanoski (1999) Approach

The resampling approach reported by Lobb and Kachanoski (1999) is based on a study carried out in Southwestern Ontario at the Brant County Study Site. The used method is based on previous contributions made by the Canadian research group (Kachanoski 1987; Kachanoski et al. 1992; Kachanoski and von Bertoldi 1996; Lobb et al. 1999) and it is built on a first set of soil samples collected in 1971 over the same site. The same locations were resampled in 1986 (after 15 years) and a new set of samples was available for ^{137}Cs analyses. On both occasions, the soil samples were collected using the same soil cores (7.6 cm diameter) to a depth of 1.5–2 m. The samples were cut at 15-cm increments before ^{137}Cs analyses. From each sampling site, two cores were taken, and the corresponding layers were mixed before ^{137}Cs analysis. The samples were analyzed using the same laboratory procedure shortly after each sampling campaign. The data reported by the authors in their original work (Lobb and Kachanoski 1999) represent the mean values of the seven transects originally established in the study area (Kachanoski et al. 1992) and, as such, they can be easily reprocessed to make some comparisons.

Based on the information reported by the authors (see Lobb and Kachanoski 1999), all the ^{137}Cs values were decay corrected to January 1986, the date of the second sampling campaign. In this respect, the authors proposed two approaches to calculate the erosion rates for the period between the two sampling campaigns (1972–1986). Both approaches follow the SRA.

The first approach, presented by de Jong et al. (1983) and defined as the 'linear method', is based on the following equation which is a variant of Eq. 2.4:

$$A = \frac{\rho_C D_C (Cs_0 - Cs_i)}{Y Cs_0} \tag{2.6}$$

where:

A is the soil erosion rate (kg m^{-2} yr^{-1});
D_C is the soil depth to which the ^{137}Cs is distributed (m);
ρ_C is the bulk density of the ^{137}Cs contaminated layer (kg m^{-3});
Y is time elapsed between the two sampling campaigns (yr);
Cs_0 is the inventory of ^{137}Cs to which Cs_i (inventory at the sampling point, i) is compared (Bq m^{-2}). More specifically, Cs_0 is the ^{137}Cs inventory related to the first campaign (corrected for decay and fallout) that would be present at the same site if no erosion had occurred.

The second approach, proposed by Walling and Quine (1991) and Kachanoski (1993), and defined as the 'power method', can be expressed by the following equation that accounts for incorporation of underlying soil with absence of ^{137}Cs into the till-layer:

$$A = \frac{\rho_C D_C \left(1 - (Cs_0/Cs_i)^{1/Y}\right)}{\eta} \tag{2.7}$$

where:

η is an enrichment ratio that, for this study site, was assumed equal to 1.

Even if Eq. 2.7 gives slightly better estimates of soil loss than Eq. 2.6, it is unable to give suitable estimates for deposition sites and, for this reason, it will not be considered here.

The application of Eq. 2.6 for 20 samples for which all the parameters are available, is carried out using the information provided by the authors about the depth of tillage (ca. 0.18 m) and the soil bulk density (ca. 1300 kg m^{-3}). The number of years (Y = 7) was established based on the fact that before the first sampling campaign the area was on grassland and in 1972 it was ploughed and used for corn for six years. After this cultivation period, the site returned to grassland with no subsequent tillage. The results of this application are plotted in Fig. 2.11.

From the results reported on Fig. 2.11, it can be deduced that, during the period between the two sampling campaigns, 11 (out of 20) sampling points showed deposition. The procedure suggests, for the 1972–1986 period, a net soil deposition of ca. 22.3 (t ha^{-1} yr^{-1}) which documents a massive soil movement over this field. However, as reported by the authors, soil erosion is concentrated on the convex upper slope positions while soil deposition occurred generally on the concave lower slopes.

Discussion: A key feature of the basic approach suggested by Lobb and Kachanoski (1999) is related to the fact that the timespan occurring between the two sets of measurements (ca. 15 years) could be long enough to provide appreciable differences between the corresponding ^{137}Cs inventories.

However, one important limitation of the approach proposed by Lobb and Kachanoski (1999) is the land use change during the studied period. The first sampling campaign was undertaken over an area that was in grassland (undisturbed land) and the authors did not report information about the ^{137}Cs distribution over a soil profile.

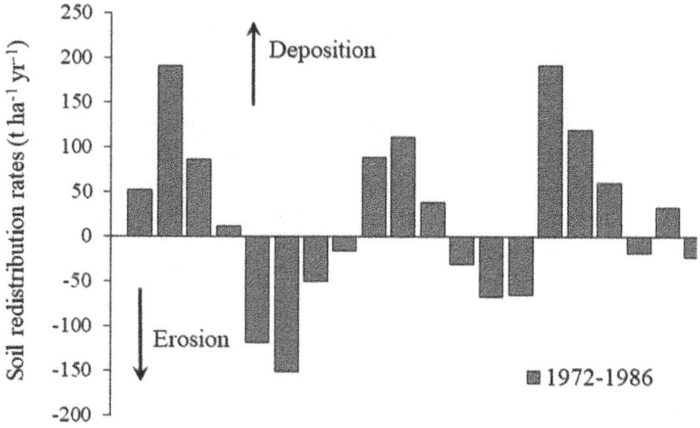

Fig. 2.11 Soil redistribution rates for the 20 sampling points derived for the 1972–1986 period using the SRA

This makes impossible to apply the technique in its traditional way for the first campaign (for example, using an exponential model) and, consequently, a comparison between the estimates obtained using the SRA and TRA cannot be made.

2.1.1.5 Tiessen et al. (2009) Approach

The resampling approach reported by Tiessen et al. (2009) is based on a study carried out in Atlantic Canada at the 20NB Study Site located near the Grand Falls town in New Brunswick. This approach is based on a previous contribution of Cao et al. (1994) and it is built on a first set of soil samples collected in 1990 over the same site. The same area was revisited in 2005 (after 15 years) and a new set of samples was collected for ^{137}Cs analyses. On both occasions 129 soil samples (45 cm deep) were collected following a grid pattern, using the same soil cores (6.2 cm diameter). The samples were cut to 15 cm depth increments before ^{137}Cs analyses. From each sampling site, two cores were taken, and the corresponding layers were bulked to create well representative composite samples for ^{137}Cs analysis. The samples were analyzed by the same laboratory procedure shortly after each sampling campaign.

During the second campaign, an undisturbed area was individuated approximately 10 km southwest of the study site to derive a ^{137}Cs reference value. In this case, 12 points were sampled using the same soil cores (6.2 cm diameter) to a depth of 60 cm. Each sample was sectioned to 15 cm depth increments. This strategy provided a ^{137}Cs reference value of 1804 (Bq m^{-2}), decay corrected to 2005. The corresponding base level of ^{137}Cs for the first sampling campaign in 1990 was estimated to be 3000 (Bq m^{-2}), from the results provided by de Jong et al. (1986).

The authors proposed two approaches to calculate the erosion rates from cesium measurements: the traditional method (TRA), that required the two reference values mentioned above; and the resampling approach using the SRA.

The traditional method is based on the following 2 equations that are used for erosional and depositional sites respectively. The first equation, proposed by Walling and Quine (1991) and Kachanoski (1993), that accounts for incorporation of underlying soil with absence of ^{137}Cs into the till-layer, was used for erosional sites:

$$A = \frac{\rho_C D_C \left(1 - (Cs_0/Cs_i)^{1/Y}\right)}{\eta} \tag{2.8}$$

where:

η is an enrichment ratio that, for this study site, was assumed to be equal to 1.

The second equation, presented by de Jong et al. (1983), was applied for depositional sites only:

$$A = \frac{\rho_C D_C (Cs_0 - Cs_i)}{Y Cs_0} \tag{2.9}$$

where:

A is the rate of soil erosion (kg m^{-2} yr^{-1});
D_C is the soil depth to which the ^{137}Cs is distributed (m);
ρ_C is the bulk density of the ^{137}Cs contaminated layer (kg m^{-3});
Y is time elapsed between the two sampling campaigns (yr);

Cs_0 is the ^{137}Cs inventory to which Cs_i (inventory at the sampling point, i) is compared (Bq m^{-2}). More specifically, Cs_0 is the ^{137}Cs inventory related to the first campaign (corrected for decay and fallout) that would be present at the same site if no erosion had occurred.

The corresponding values of soil erosion and deposition, expressed in t ha^{-1} yr^{-1}, and related to 65 sampling sites are reported by the authors in their Table 3 (Tiessen et al. 2009). These values are replotted in Fig. 2.12.

It is clear from Fig. 2.12 that some of the points that showed deposition for the 1960–1990 period changed to erosion for the 1960–2005 period and vice-versa. Also, the sampling points from the first campaign in 1990 showed higher erosion rates (mean of 26.9 t ha^{-1} yr^{-1}) than those obtained for the second campaign in 2005 (mean of 23.4 t ha^{-1} yr^{-1}). The overall results suggest that a general deposition would have occurred during the period between the two sampling campaigns (1991–2005). The authors documented the soil redistribution rates for the 1991–2005 period in their Table 3. These values, that were determined using the SRA, are replotted in Fig. 2.13.

These results suggest that, during the period between the two sampling campaigns, the movement of soil within the sampling sites was considerable. The mean net erosion obtained for this period, using the SRA, was 13.5 t ha^{-1} yr^{-1}. Even though

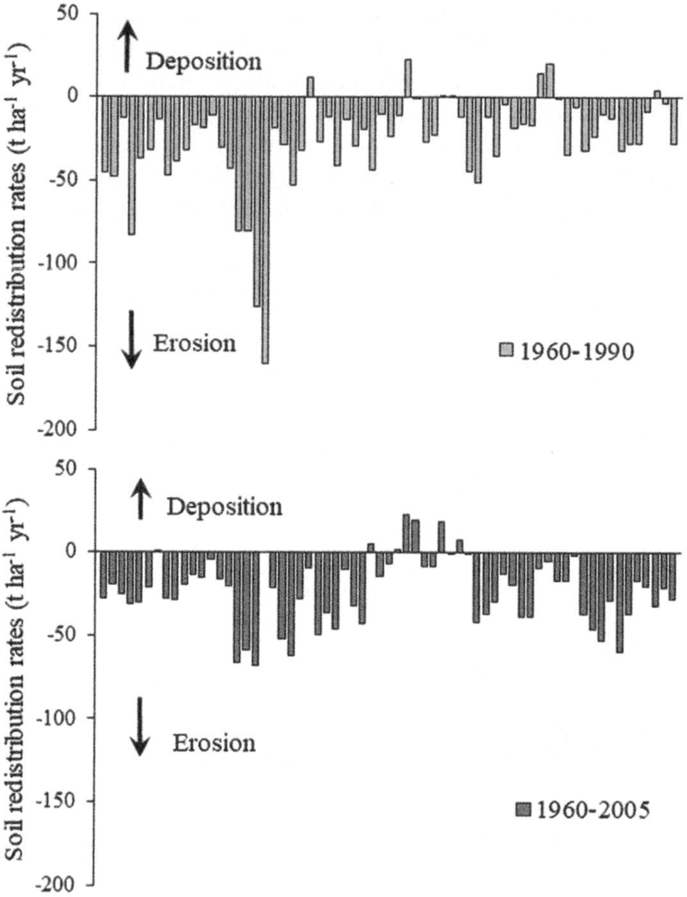

Fig. 2.12 Comparison between the estimates of soil redistribution rates for 65 sampling points for the 1960–1990 and 1960–2005 periods using the traditional approach (replotted from Tiessen et al. 2009)

this value was half that obtained using the traditional approach for both sampling campaigns, it was more than double that obtained in the study site during the 1990–2005 period from field measurements (mean value from runoff measurements of 5.3 t ha^{-1} yr^{-1}). However, the values of soil loss measured during the 1990–2005 period, obtained with an H-flume monitoring system placed downstream the study area, can not be directly compared with the estimates obtained with the ^{137}Cs measurements because this device worked only during the growing season (from May to October) and this choice may have included uncertainty in the annual values. Also, the 20NB Study Site is not a closed system and the direct observations of soil loss obtained with the H-flume are not related to the same area covered by the ^{137}Cs sampling points. The authors pointed out this problem and explained that the discrepancy

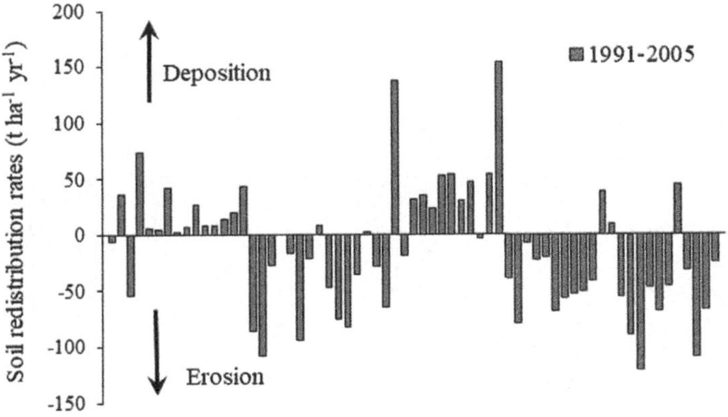

Fig. 2.13 Soil redistribution rates for 65 sampling points derived for the 1991–2005 period using the SRA (replotted from Tiessen et al. 2009)

between soil loss derived from ^{137}Cs measurements and direct observations is due to several reasons that include the sampling device, the contribution of tillage and, most importantly, the fact that the measured runoff data are related to only a portion of the entire field sampled for ^{137}Cs analysis.

Discussion: A key feature of the basic approach suggested by Tiessen et al. (2009) is related to the fact that the timespan occurring between the two sets of measurements (ca. 15 years) could be long enough to provide detectable differences between the corresponding ^{137}Cs inventories, especially because during the time span between the sampling campaigns no additional ^{137}Cs fallout occurred in the study area. However, when these inventories are plotted against the 1:1 line for comparison (see Fig. 2.14), it is clear, that the 1990 inventories are of the same magnitude than the corresponding values related to the sampling undertaken in 2005 (decay-corrected to 1990).

Two important messages resulted from these data:

Fig. 2.14 Comparison of the corresponding ^{137}Cs inventories for the same sampling sites related to the sampling campaigns in 1990 and 2005. The 2005 inventories are decay corrected to 1990

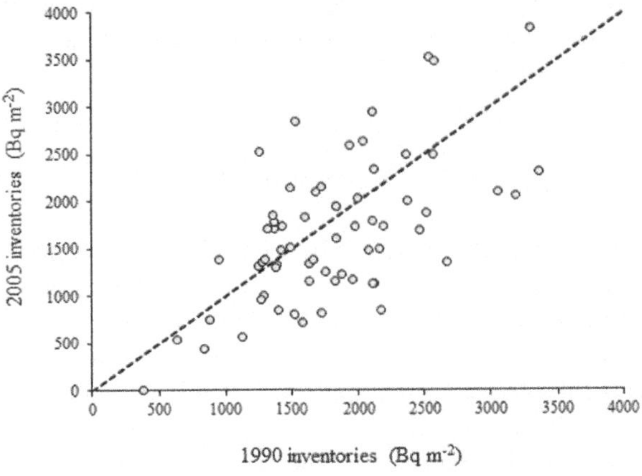

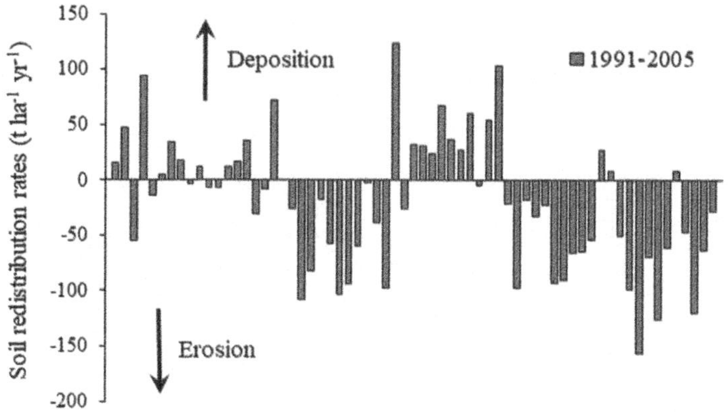

Fig. 2.15 Soil redistribution rates for 65 sampling points derived for the 1991–2005 period using the TRA (data from Tiessen et al. 2009)

1. It seems, in contrary to the previous hypothesis, that the time span occurring between the two sampling campaigns may not be long enough for erosion to make significant impact;
2. New hypothesis can be proposed that the uncertainty related to the inventories corresponding to the same sampling points is high and should be considered when comparing the ^{137}Cs inventories representing two different time horizons. This issue will be discussed more in detail in Subchapter 2.2.

However, the authors report all the measurements for the two sampling campaigns and this makes possible to apply the TRA for comparison with the SRA. The results obtained using the TRA are reported on Fig. 2.15 and can be compared with the corresponding results obtained using the SRA and illustrated on Fig. 2.13.

Data plotted on Fig. 2.15 show a similar pattern of soil redistribution rates to those calculated by the authors using the SRA. However, the mean value of net erosion derived from the TRA (19.3 t ha^{-1} yr^{-1}) is higher than that obtained with the SRA (13.5 t ha^{-1} yr^{-1}) and much higher than the measured value from the field sampling device. The overall results seem to suggest that the TRA is less accurate than the SRA at this site.

2.1.1.6 Li et al. (2011) Approach

The resampling approach by Li et al. (2011) is the last contribution from Canada. This paper represents an update of the work by Lobb et al. (1999), already discussed previously, and it is based on the same measurements carried out in the Brant County Site (Ontario). The study area and the measurements (sampling campaigns in 1971 and 1985) have been already described in the above paragraph. In this contribution, the authors pointed out on the main differences between the single sampling and the resampling approach. The single sampling approach was applied using a ^{137}Cs reference inventory value obtained as a mean value of the on-site measurements obtained in 180 single sampling points established in 1971. To use this value as a reference value for the measurements undertaken during the second sampling in 1985, this value was decay corrected to January 1986 and adjusted to account for the ^{137}Cs fallout between 1972 and 1985. The authors applied also the traditional approach using a biased reference value (ca. 10% chance of bias) to explore the effect of bias included in the estimate of a hypothetical off-site reference value. These two versions of traditional approach have been compared with the application of the SRA based on the two sampling campaigns undertaken in 1971 and 1985. The overall results suggested that the traditional approach, if the reference value is accurate, provides a better explanation of the pattern of soil redistribution rates obtained after conversion. However, if the reference value is biased due to local and/ or regional variability of ^{137}Cs fallout, even the soil redistribution rates will result biased, and this bias cannot be reduced by taking multiple samples. On the contrary, if the resampling strategy is applied, the errors contained in the choice of the reference value may be eliminated and the errors due to the random uncertainty can be reduced by taking multiple samples.

Discussion: The paper reported by Li et al. (2011) is certainly the first contribution that tries to compare the use of single sampling and resampling approach in soil erosion estimation and pointed out some key issues on the application of ^{137}Cs technique.

The first issue relates to the use of a single reference value when applying the traditional approach. As the authors suggested, this value, especially if based on a few measurements, can be biased. This is a correct assumption but, if the number of samples increases, the uncertainty related to this estimate can be reduced and, consequently, the estimates of soil redistribution rates are more accurate, especially if the reference site is close to the study area (Sutherland 1996). Unfortunately, the authors did not select an off-site reference area in their work, and it is not possible to explore further this uncertainty.

A second key issue is related to the resampling technique and focusses on the random errors linked to the single sampling values. Based on the assumptions made by the SRA, this is a problem that occurs when the inventories of the same sampling site are compared. Although the two inventories to be compared (one for each sampling campaign) are attributed to the same point, this is an artificial imagination because it is impossible to resample exactly the same location twice. The suggestion given by the authors to collect multiple samples and to use the spatially averaged erosion rates for data interpretation is a good solution to reduce this uncertainty and should be considered when the resampling strategy is applied.

Another key issue related to this contribution is the absence of a reference sampling for the 1971 sampling campaign. This makes the application of the TRA impossible and, consequently, no comparison between resampling methods can be done.

For assessing the results of single sampling and resampling approaches it would be very useful to compare both approaches to third data independent from use of FRN tracers, for example measurements at erosion plots or measurements of suspended sediment yields at elementary watershed. Unfortunately, such data are not available not only for this study but for a great majority of resampling studies conducted up to now. For future it should be highly recommended to design such studies which would enable validating FRN methods in general and comparison of single sampling and resampling approach of FRN methods.

2.1.2 Contributions from the USA

The pioneering studies using ^{137}Cs for soil erosion assessment were conducted in the USA by J. McHenry and J. Ritchie during the early 1970s (Ritchie et al. 1970, 1974; McHenry et al. 1973). These works were based on the application of the ^{137}Cs technique in its traditional way and their main goal was to focus on the advantages offered by this technique over the conventional methods of soil erosion measurement, such as erosion pins and wash traps. Several years later, some of these study areas (Fig. 2.16) were resampled, and the existing datasets were integrated with new measurements to detect possible change in soil erosion rates indicating their development trends. These attempts reported by McHenry and Bubenzer (1985) and Fornes et al. (2005) are discussed in the following two subchapters.

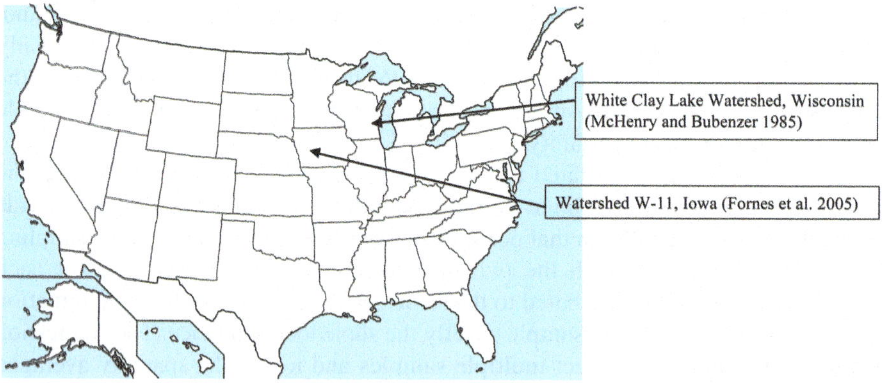

Fig. 2.16 Experimental sites in USA that have been resampled for ^{137}Cs analysis by McHenry and Bubenzer (1985) and Fornes et al. (2005)

2.1.2.1 McHenry and Bubenzer (1985) Approach

The study reported by McHenry and Bubenzer (1985) was probably the first attempt in the USA to employ the resampling method associated with the use of ^{137}Cs to estimate soil erosion. The method is based on previous contributions made by the American research group in Wisconsin (McHenry et al. 1973; McHenry and Ritchie, 1977; Mitchell et al. 1980) and it is built on a first set of soil samples collected in 1974 and 1976 from cultivated areas in the White Clay Lake Watershed (see Fig. 2.16). The same sites were resampled in 1980 (after 4–6 years). At both occasions, the soil samples were collected at 5 cm depth increments to a depths of 30 cm and at 10 cm increments from 30 to 60 cm. From each sampling site, six cores were taken, and the corresponding layers were bulked to obtain well representative composite sample for ^{137}Cs analyses. During the same campaigns, a wooded, undisturbed area in the vicinity of the study sites was selected to derive a ^{137}Cs reference value. In this case, eight soil profiles were sampled in 1974 under a mixed oak-maple forest and two depth incremental cores were collected in 1980 from the same site. The corresponding reference inventory was 4572 (Bq m^{-2}) for 1974, and 3885 (Bq m^{-2}) for 1980.

The samples collected in the study area and at the reference site were analyzed using the same laboratory procedure shortly after each sampling campaign. The single ^{137}Cs inventories obtained for both campaigns were compared with the corresponding base level and estimates of soil erosion and/or deposition were obtained for the 1954–1974 and 1954–1980 periods assuming that the commencement of ^{137}Cs fallout was in 1954. Even if the authors state that a relationship of proportionality between cesium loss/gain and soil loss/deposition was assumed, it is not clear which conversion model was used. However, in their original contribution McHenry and Bubenzer (1985) provide, in graphical forms, the values of soil erosion and deposition expressed in cm. These values related to the three longest transects established in the Brunner Farm are replotted on Fig. 2.17 recalculated to t ha^{-1} yr^{-1} assuming a bulk density of 1000 kg m^{-3}.

From Fig. 2.17 it is clear that some deposition points of the 1954–1974 period showed erosion for the 1954–1980 period and vice-versa. Also, some of the points that showed high values of soil erosion (or deposition) in the 1954–1974 period showed lower rates of erosion (or deposition) in the 1954–1980 period and vice-versa.

The authors reported a summary of their calculations in their Table 2 (McHenry and Bubenzer 1985). In this table, for the 1954–1976 period, soil erosion rates proved to be dominant for the 3 transects investigated. These values range from 1.84 t ha^{-1} yr^{-1} to 24 t ha^{-1} yr^{-1}. On the contrary, for the 1954–1980 period, net deposition rates were obtained, and these values range from 2.02 t ha^{-1} yr^{-1} to 20.4 t ha^{-1} yr^{-1}. These results produced an estimate of net deposition for the 1976–1980 period ranging from 83.8 to 135 t ha^{-1} yr^{-1}. It is not clear how these estimates were provided because the authors did not report details of their calculations and of the conversion model used. However, the overall results suggest that, during the period between the two sampling campaigns, the movement of soil within the sampling sites was considerable. In this respect, the authors deduced that these estimates of

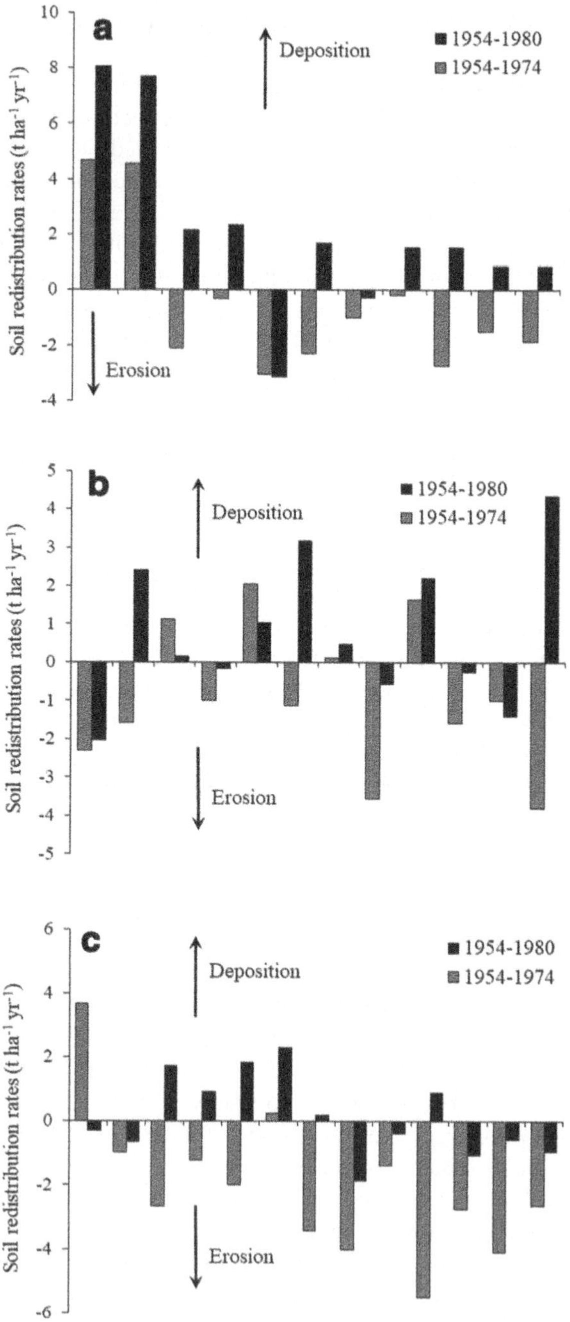

Fig. 2.17 Comparing soil loss for the 1954–1974 and 1954–1980 periods related to transects established at (**a**) the Brunner Farm Field 5, (**b**) the Brunner Farm Field 4, and (**c**) the Brunner Farm Field 221-121-115 (replotted from McHenry and Bubenzer 1985)

net deposition rate are attributed to the topographic position of each sampled transect. In fact, because the sampling sites are located downstream of other cultivated areas, the sampling points have received sediment contribution from the fields located upstream. To prove the evidence of that, the estimates of soil loss from the upstream areas were provided using the Universal Soil Loss Equation (USLE). They ranged from 1.3 (t ha^{-1} yr^{-1}) to 4.0 (t ha^{-1} yr^{-1}) what is much lower than the deposition rates provided by the ^{137}Cs measurements (McHenry and Bubenzer 1985).

Discussion: A key feature of the approach proposed by McHenry and Bubenzer (1985) is that the value of ^{137}Cs inventory obtained for each sampling point is compared with the relevant reference value obtained in an undisturbed area using the traditional approach (TRA). Following this strategy, a mean value of soil redistribution rate for the two periods of 1954–1974 and 1954–1980 can be obtained for each sampling point (based on the figures provided by the authors) and it is possible to apply the TRA to integrate over time these estimates to derive the soil redistribution rates for the 1975–1980 period only. The reference value obtained in 1980 is well in agreement with that obtained in 1974 and this indicates a significant reliability of this approach. This is better emphasized by Fig. 2.18 related to the 1975–1980 period only.

The values of soil redistribution rate presented on Fig. 2.18 were obtained using the TRA following these steps:

1. The estimates (t ha^{-1} yr^{-1}) obtained for the 1974 sampling campaign were time integrated (t ha^{-1}) for the 1954–1974 period, assuming that the commencement of ^{137}Cs fallout was in 1954.
2. Accordingly, the estimates (t ha^{-1} yr^{-1}) obtained for the 1980 sampling campaign were time integrated (t ha^{-1}) for the 1954–1980 period.
3. The point-by-point difference between the corresponding estimates obtained above provided the cumulative soil redistribution (t ha^{-1}) for the 1975–1980 period.
4. These values were then divided by the number of years (6, for this specific case) elapsed between the two sampling campaigns and the final estimates of soil redistribution rates (t ha^{-1} yr^{-1}) were obtained.
5. The results showed on Fig. 2.18 indicate again, that a massive movement of soil occurred during the time window since the first sampling until the resampling. Based on our calculations and the assumptions made, a net gain of soil is documented for each transect. These estimates are ca. 10.94 (t ha^{-1} yr^{-1}) for the Brunner Farm Field 5, ca. 6.45 (t ha^{-1} yr^{-1}) for the Brunner Farm Field 4, and ca. 7.57 (t ha^{-1} yr^{-1}) for the Brunner Farm Field 221-121-115. Our results, even if different from the estimates provided by the authors, confirm that a net deposition was the dominant process within the area during the period between the first and the second sampling.
6. There are however some limitations regarding the results obtained using this approach. The first limitation is related to the fact that the time window between the two sampling campaigns is short (only six years) and this may incorporate some uncertainty in the estimates of soil redistribution rate. In this respect, the

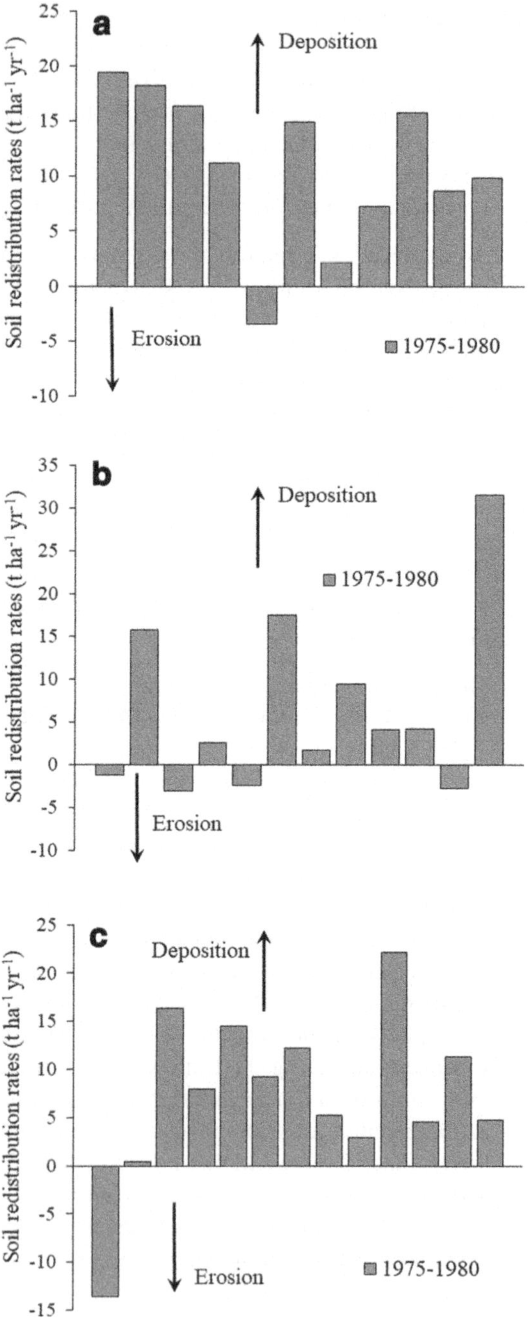

Fig. 2.18 Erosion and deposition rates for the 1975–1980 period related to transects established at (**a**) the Brunner Farm Field 5, (**b**) the Brunner Farm Field 4, and (**c**) the Brunner Farm Field 221-121-115

uncertainty related to the small-scale variability of ^{137}Cs due to its heterogeneous initial fallout and bioturbation can be of the same order of magnitude of the variation of inventories due to erosion and deposition. In their previous work, McHenry and Ritchie (1977) reported the ^{137}Cs activities for the eight soil profiles collected at the reference site. These profiles are typical for undisturbed sites, but they show a CV of ca. 13.4% due to the microscale variability. Even if this value is in the acceptable range suggested by other authors (Sutherland 1996) to make the cesium technique suitable, the reference inventory variation between 1974 and 1980 (as reported by the authors) is ca. 13.5%, which is surprisingly close to the former. In other words, it is difficult to establish to which extent the uncertainty related to the reference values has affected the final estimates obtained by the 1980 sampling campaign and this makes the suitability of the results obtained by the authors questionable. Unfortunately, the authors did not report the ^{137}Cs inventories for the second sampling campaign and this made the application of the SRA not possible and precluded the comparison with the above results provided by the TRA. Another limitation is related to the fact that independent measurements of soil loss are available only for the areas downstream of the study sites and, even if some USLE estimates are provided, it is difficult to establish if the approach provides suitable predictions of soil erosion rates for that area.

2.1.2.2 Fornes et al. (2005) Approach

The resampling approach proposed by Fornes et al. (2005) is based on a previous study made by Spomer et al. (1985) in the watershed W11, located in Iowa (see Fig. 2.16). Spomer et al. (1985) established a downslope transect in a cultivated catchment, comprising of 9 sampling points and in 1974 it was sampled for ^{137}Cs measurements. In 1998 (24 years after the first sampling), the same transect was resampled by Fornes et al. (2005) and the samples were taken as close as possible to the original points sampled in 1974. In 1974 the sampling depth was 35 cm by 5 cm depth increments for both the transect and the reference area. In the reference area three depth incremental cores were collected for determining the initial concentration of ^{137}Cs fallout. During the sampling campaign in 1998, soil cores of 2.5 cm diameter were collected to a depth of 60 cm by 5 cm increments. In the reference area, two sectioned cores were collected using the same sampling strategy. The authors developed a new model (CSEM—^{137}Cs Soil Erosion Model) to calculate the erosion rates for the periods of 1954–1974, 1954–1998, and 1975–1998. CSEM is a time-dependent, multi-box model that is based on the following mass balance equation:

$$\frac{\partial I}{\partial t} = F(t) - EC(t) + \lambda(I) \tag{2.10}$$

where:

t is time (yr);
I is the total ^{137}Cs inventory (Bq m^{-2});
E is the rate of soil erosion (g m^{-2}t^{-1});
$C(t)$ is the time-dependent ^{137}Cs activity in plough layer (Bq g^{-1});
$F(t)$ is the fallout input (Bq m^{-2});
λ is the ^{137}Cs radioactive decay (t^{-1}).

The model requires apart of the measurements of ^{137}Cs inventory for each sampling site also the values of ^{137}Cs fallout. Because no measurements of atmospheric fallout were available in the study area, the authors used the data available at the Argonne National Laboratory (Illinois, USA) and Milford Haven (UK), adjusting the magnitude based on the local inventories. The reference sampling was done in undisturbed area located in the vicinity of the study sites. A mean value of 5712 Bq m^{-2} was obtained in 1974 and 2947 Bq m^{-2} was deduced for 1998. The authors calculated the erosion rates for the 1954–1974 and 1954–1998 periods. In this respect, they applied their model using the traditional assumptions required by the ^{137}Cs technique where the 9 single ^{137}Cs inventories, for each sampling campaign, were compared with the corresponding reference value. Then, to derive the erosion rates for the period between the two sampling campaigns (1975–1998), the two resampling interpretation approaches (TRA and SRA) were used:

The TRA was applied by subtracting the time weighted contribution of the erosion rate calculated for the 1954–1974 period from the contribution of the erosion rate calculated for the 1954–1998 period;

The SRA was applied assuming as reference value the ^{137}Cs inventory obtained during the first campaign for each corresponding point.

The corresponding values of soil erosion and deposition rates, expressed in g cm^{-2} yr^{-1}, are reported by the authors in graphical form (Fornes et al. 2005). These values are replotted on Fig. 2.19 in t ha^{-1} yr^{-1}.

Discussion: Based on the information shown on Fig. 2.19, the authors deduced that the estimates of soil erosion for the 1954–1974 period were significantly higher

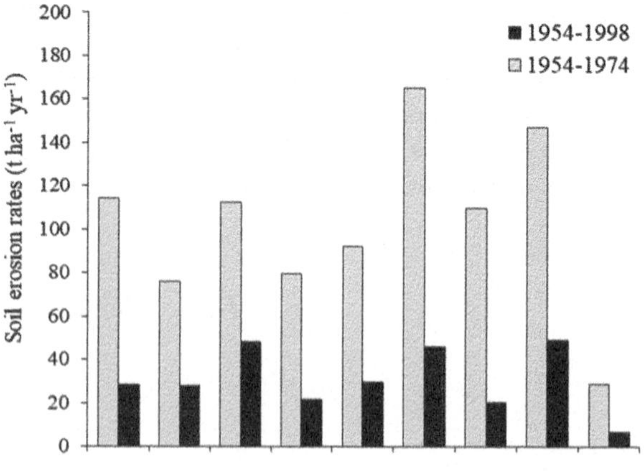

Fig. 2.19 Comparison between soil loss for 9 sampling points for the 1954–1974 and 1954–1998 periods. (replotted from Fornes et al. 2005)

than those related to the extended 1954–1998 period. The reason of this reduction in soil erosion rates was attributed to the improved agricultural practices used since 1991. Some important considerations were made on the use of the two methods to derive the rates of soil erosion during the nonoverlapping period (1975–1998). More specifically, the use of TRA indicated net deposition over the period between the two sampling campaigns. These results were in contrary with the field data of soil loss measured at a weir station located at the catchment outlet.

On the contrary, the SRA indicated that the mean soil loss during that period was ca. 19 t ha^{-1} yr^{-1}, in reasonable agreement with that provided by the weir station (17.7 t ha^{-1} yr^{-1}). The authors concluded that the TRA was inappropriate for their location.

2.1.3 Contribution from Australia

The pioneering studies involving the use of ^{137}Cs in Australia were conducted during the early 1980s (Campbell et al. 1982, 1986; Longmore et al. 1983). These and the subsequent works (Elliott et al. 1990; Loughran et al. 2004) were based on the application of the ^{137}Cs technique in its traditional form. Several years later, one of these study areas, located in the Hunter Valley (Fig. 2.20) was resampled, and the existing dataset (Campbell et al. 1986) was integrated with new ^{137}Cs measurements in order to detect possible change in soil erosion rates for a 19-year period (see Loughran and Balog 2006).

The results of this study, based on cultivated soils, are synthesized and discussed in the following section.

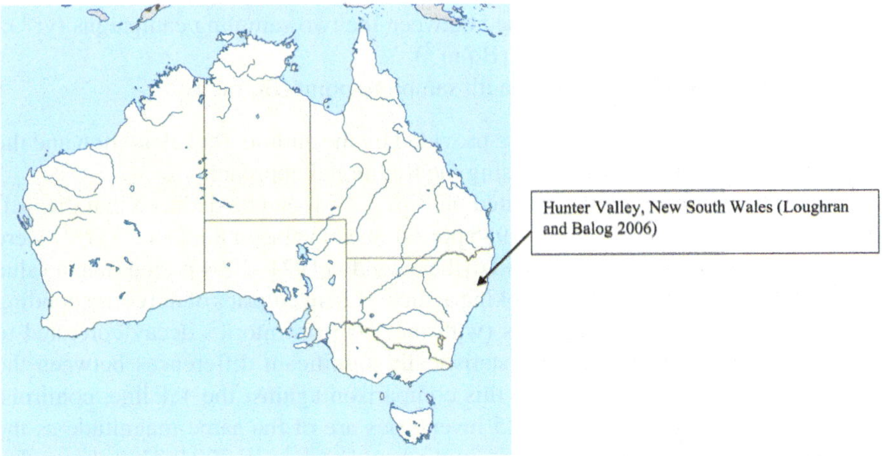

Fig. 2.20 Experimental sites in Australia that have been resampled for ^{137}Cs analysis by Loughran and Balog (2006)

2.1.3.1 Loughran and Balog (2006) Approach

The resampling approach proposed by the authors is based on a previous study made by Campbell et al. (1986) in the Maluna Vineyards, Hunter Valley, located in New South Wales. In their original work, Campbell et al. (1986) established a grid-based sampling, where 27 points were sampled in 1984–1985 for ^{137}Cs measurements. In 2004 (ca. 19 years after the first sampling), the same site was resampled by Loughran and Balog (2006) and 29 samples were collected as close as possible (ca. 5 m) to the points sampled in 1984. Soil collection in 1984 was made to a depth of 20 cm using a steel core with an internal diameter of 9.5 cm. For the second sampling a slightly different steel core (with an internal diameter of 8.5 cm) was employed, and the sampling depth remained the same (20 cm). Information on the ^{137}Cs reference value (1040 Bq m^{-2}) is available only for the first campaign in 1984 during which a hilltop above the vineyard was sampled using a scraper plate (20 cm × 50 cm) to determine the ^{137}Cs depth distribution and the total ^{137}Cs inventory (Campbell et al. 1986). The reference value for the second sampling (678 Bq m^{-2}) was derived by correcting the value from 1984 for the decay.

To convert loss and gain of ^{137}Cs into soil erosion and deposition, the authors employed a simple proportional model based on the following equation:

$$A = \frac{10B\,d\left(Cs_{Ref} - Cs_i\right)}{100\,T\,Cs_{Ref}} \tag{2.11}$$

where:

A	is the rate of soil erosion (t ha^{-1} yr^{-1});
d	is the soil depth of cultivation (here assumed to 13 cm);
B	is the soil bulk density (kg m^{-3});
T	is the number of years elapsed between the two sampling campaigns (yr^{-1});
Cs_{Ref}	is the ^{137}Cs reference value (Bq m^{-2});
Cs_i	is the ^{137}Cs inventory for the ith sampling point (Bq m^{-2}).

All the parameters of Eq. 2.11 are provided by the authors for calculation and the results are shown on Fig. 2.21 by using the traditional approach.

Based on the information presented on Fig. 2.21, it can be deduced that the estimates of soil erosion for the 1971–1985 period (mean value of 80.2 t ha^{-1} yr^{-1}) were significantly higher than those related to the extended 1971–2004 period (mean value of 34.2 t ha^{-1} yr^{-1}). The authors, looking at the ^{137}Cs inventories of the corresponding points of both sampling campaigns (with the 1985 inventories decay corrected to 2004), observed that there are no statistically significant differences between the two datasets. Figure 2.22 showing this comparison against the 1:1 line, confirms, with some exceptions, that the 1985 inventories are of the same magnitude as the corresponding values related to the sampling undertaken in 2004. This shows that the net erosion values during the second time window (1986–2004) are much lower than during the first time window (1971–1985). This difference could be caused by possible changes in rainfall erosivity (as documented by the authors in their original work), land management (after 1971 a vineyard was established but the commencement of sod-culture in 1998 has increased groundcover and reduced the impact of rainfall and runoff erosivity), and differences in sampling and laboratory methods.

Discussion: A key feature of the approach proposed by Loughran and Balog is related to the fact that they reported all the measurements obtained during the two sampling campaigns, for each of which they applied the ^{137}Cs technique following traditional method. For this reason, it is possible to derive the erosion rates for the second time window (1986–2004), using both resampling interpretation approaches (TRA and SRA). These calculations were made as follows:

The TRA was applied by subtracting the time weighted contribution of the erosion rate calculated for the 1971–1985 period from the contribution of the erosion rate calculated for the 1971–2004 period.

The SRA was employed assuming as reference value for the 2004 sampling the ^{137}Cs inventory obtained during the first sampling for each corresponding point.

The results of this application are compared on Fig. 2.23a, b.

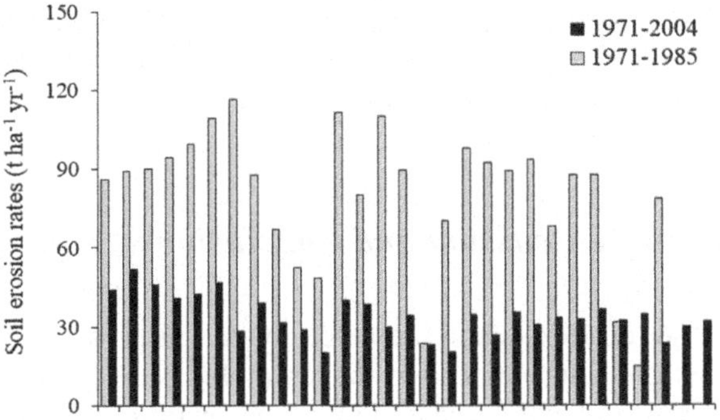

Fig. 2.21 Comparison between the soil loss for the 29 sampling points for the 1971–1985 and 1971–2004 periods (replotted from Loughran and Balog 2006)

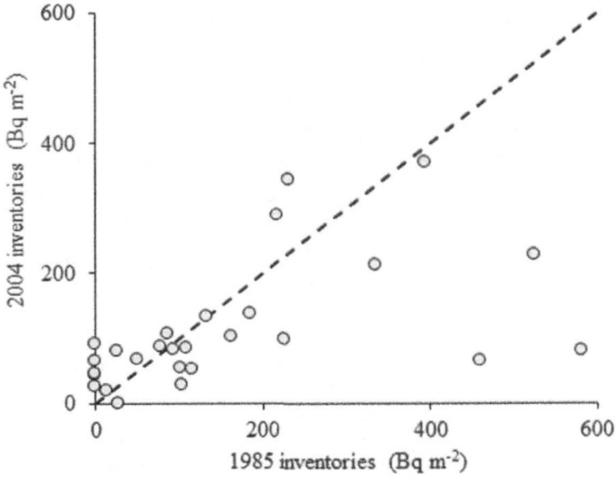

Fig. 2.22 Comparison of the ^{137}Cs inventories of the same sampling points sampled in 1985 and 2004; The 1985 inventories are decay corrected to 2004

Figure 2.23 clearly demonstrates that, if the ^{137}Cs resampling is carried out for each single point, a massive movement of soil is recorded during the second time window. Based on the rough calculation and the assumptions made, a net soil loss of ca. 0.45 t ha^{-1} yr^{-1} is obtained if we use TRA, while a net deposition of ca. 8.7 t ha^{-1} yr^{-1} is obtained if the SRA is employed. Even though both approaches (TRA and SRA) suggest a noticeable soil movement during the recent period (1985–2004), the patterns in Fig. 2.23a,b and the related estimates are different, and it is difficult to find an explanation for this discrepancy. Direct measurements of soil erosion are not available for this site, so it is not possible to compare both resampling approaches with another independent dataset to find out which approach is more correct. However, it is worth noticing that the authors treated the measurements from the two sampling campaigns as two independent datasets. Following this approach, they applied the TRA on the mean values of the two datasets. Based on their calculation, a value of net soil loss of ca. 9.7 t ha^{-1} yr^{-1} was obtained. This value is different from that (0.45 t ha^{-1} yr^{-1}) obtained by our calculations and we think that the former is not correct if the data reported by the authors in their Table 1 are correct. Following the same calculation strategy, if the SRA is applied (using the mean values of the two datasets) a net soil loss of ca. 25 t ha^{-1} yr^{-1} is obtained. Again, this value is different from that (net deposition of 8.7 t ha^{-1} yr^{-1}) obtained by our calculations. This is not surprising because our calculations using SRA are based on a 'point by point' comparison and this has affected the results.

2.1.4 Discussion and Conclusions

Based on the examples reported above, it is possible to outline some general remarks on the results obtained by the resampling strategy in English speaking countries and to give some preliminary indications on how the two approaches can be used and to which extent they can be considered reliable in soil erosion investigations.

1. A first point of discussion is why in several occasions the TRA gave different soil erosion rates than the SRA. The only exception is the work of de Jong and Kachanoski (1988) where both approaches provided similar results. This is not surprising because the two approaches are based on different assumptions. The results obtained with the SRA are strongly affected by the values of the previous inventories measured at the corresponding sampling points. This approach does not provide data for the first time window (since the ^{137}Cs fallout until the first sampling) but it provides results only for the second time window (since the first sampling until the second sampling). If the microscale variability related to each single point is high, the resulting estimates can be biased. However, the great advantage of this approach is that there is no need of separated reference site and reference value, and this avoids the uncertainty related to the base level of ^{137}Cs in the area of interest. In contrary, the TRA requires the two reference values (one for each sampling campaign) and this can be a problem if no fallout

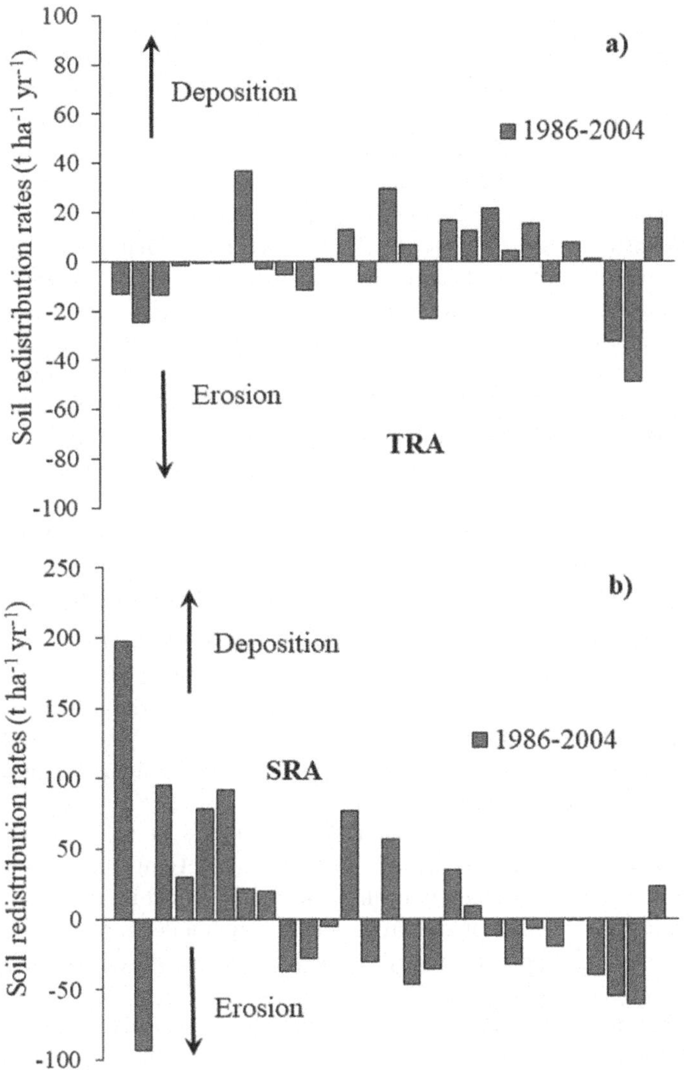

Fig. 2.23 Comparison between the estimates of soil redistribution rates for the 29 sampling points for the 1986–2004 period; The estimates are related to the application of the TRA (a) and SRA (b)

measurements are available and if the reference value is biased also the calculated erosion rates would be biased. However, the TRA has the great advantage that it explores two time windows (since the ^{137}Cs fallout until the first sampling and since the ^{137}Cs fallout until the second sampling) and, as such, it takes into account the processes occurred in the past (i.e. prior to the first sampling). This strategy allows to treat the two datasets of soil erosion rates as independent datasets, and this offers some advantages when the estimates are compared.

2. A second important indication is that when independent measurements of soil erosion are available for comparison, the SRA gave a better performance than the TRA. This is well confirmed by our further elaboration of the data provided by Kachanoski (1987) and Tiessen et al. (2009). More particularly, the contribution of Kachanoski (1987) is the only example where the comparison is made for overlapping periods and the scale of the experiment (small erosion plots) gives more strength to this conclusion. This is also documented by Fornes et al. (2005) who pointed out the different estimates given by the SRA and TRA and suggested that the latter is not appropriate for their study area. This is a difficult point to be discussed as the contemporary information on measurements of soil erosion and estimates provided by ^{137}Cs is available only for Canada (see Kachanoski 1987) and is limited to cultivated soils. This issue will be discussed in Subchapter 2.2 of this book where three case studies involving both measurements of sediment yield and erosion estimates obtained by ^{137}Cs in uncultivated soils are compared.
3. In some cases, the application of the resampling technique documented massive movements of soil (Lobb and Kachanoski 1999; McHenry and Bubenzer 1985; Loughran and Balog 2006). This can be explained by several factors:

    a. One of the main reasons can be attributed to the short timespan between sampling campaigns (see the work of McHenry and Bubenzer 1985). In this respect, the uncertainty related to the small-scale variability of ^{137}Cs due to its heterogeneous initial fallout can be of the same order of magnitude as the inventory variation in the reference area and this affects the final estimates of soil redistribution rate.
    b. Another reason can depend on the inaccuracy of the conversion model employed. For example, in the contribution of McHenry and Bubenzer (1985) no details are provided about the conversion model used by the authors to obtain the estimates of soil erosion rate. This model is probably based on a relationship of proportionality between ^{137}Cs loss and soil loss and this may have affected the final results.
    c. Problems due to microscale uncertainty are possible. Loughran and Balog (2006) pointed out this issue but in their case the problem should be limited as their work is based on cultivated land where soil is mixed over the ploughed horizon by tillage operations. This reduces the random variability due to non-uniform fallout, bioturbation and disturbance caused by the previous sampling campaign. However, if the SRA is applied using a point-by-point approach instead of a comparison between the means of the two datasets some discrepancies in the final results are evident. This difference, in our opinion, can be explained by their very low ^{137}Cs inventories with most of them very close to the detection limit. This would have affected the values of each single sampling point and this uncertainty would be propagated on the entire area. In similar cases, using the comparison between mean values, as the authors did, may reduce this uncertainty. This concept was recalled by

Li et al. (2011) who suggested the use of mean inventory values (grouping the single sampling points into landform elements) to reduce this uncertainty and this can be a useful solution to the problem.

4. The last point of discussion is the impact of tillage. All examples of resampling studies reported in this Chapter represent cultivated land. As discussed above this gives some advantages for the use of resampling technique because at cultivated land the microvariability is limited because of periodic mixing of soil by tillage operations. When uncultivated soils are under investigation, this effect of microvariability is much higher. In fact, ^{137}Cs distribution in uncultivated soils is generally exponentially decreasing with soil depth and any small disturbance of the uppermost soil layer (for example by bioturbation) may result in strong differences in the total ^{137}Cs inventory and, consequently, strong deviation in the estimates of erosion rates. Such conditions are evaluated in subchapter 2.2 reporting three case studies from uncultivated small catchments. For each of these case studies, only the TRA was applied to treat each dataset (corresponding to each sampling campaign) independently. The possible application of the SRA requires further investigation and additional work to be proposed as a general resampling approach for uncultivated soils.

Acknowledgments The authors would like to extend their gratitude and acknowledgements to all authors and study team members of the original publications cited in this chapter for their time and energy spent on providing the original papers and the related information useful to produce this chapter. These are as follows: for the contributions from Canada: Begg, C.B.M., Bettany, J.R, Chow, T.L., de Jong , E., Gregorich, E.G., Kachanoski, R.G., Li, S., Lobb, D.A., McConkey, B.G., Mehuys, G.R., Miller, M.H., Protz, R.D., Rees, H.W., Tiessen, K.H.D., Villar, H., & von Bertoldi, A.P.; for the contributions from the USA: Bubenzer, G.D., Fornes, W.L., McHenry, J.R., Matisoff, G., Piest, R.F., Spomer, R.G., Whiting, P.J., & Wilson, C.G.; for the contributions from Australia: Balog, R.M., & Loughran, R.J.

2.2 Using ^{137}Cs Resampling Approach in Europe

In Europe, pioneering studies exploring the use of ^{137}Cs in soil erosion investigations were conducted during the early 1980s in Italy (Bazzoffi and Panicucci 1983). During the following years, UK researchers (for example Walling et al. 1986) promoted the use of ^{137}Cs as a tool for documenting soil erosion and deposition in different areas of Europe. Much of this work focused on understanding ^{137}Cs behavior in different soils (undisturbed and cultivated) and developing models for deriving estimates of soil redistribution rates from ^{137}Cs measurements. This provided the basis for using the ^{137}Cs technique in its traditional form to estimate soil redistribution rates in different environments (Zhang et al. 1990a; Walling and Quine 1991; He and Walling 1997; Walling and He 1999).

During the following decades, many studies, based on the pioneering works mentioned above used the ^{137}Cs technique in its traditional form to investigate soil erosion and deposition rates in different countries of Europe. Examples are available from many countries that include Austria (Mabit et al. 2009), France (Sogon et al. 1999), Germany (Kirchner and Ehlers 1998), Italy (Porto et al. 2001), Poland (Higgitt et al. 1992), Romania (Ionita and Margineanu 2000), Russia (Sidorchuk and Golosov 1996), Slovakia (Fulajtar 2000), Slovenia (Zupanc and Mabit 2010), Spain (Navas et al. 2005), Switzerland (Konz et al. 2009), and The Netherlands (de Roo 1991). However, only recently, the ^{137}Cs resampling approach was tested in Europe and its application is limited to Italy and Spain.

In this Chapter, the experience gained in these two countries, will be described. The first subchapter (2.2.1 Contributions from Italy) presents three examples from Italy. First two examples are related to two experimental watersheds (RLW and EOW) located in the experimental area of *Crotone* (Calabria) which have already been published (Porto et al. 2014, 2018). These results will be discussed based on new information related to these watersheds and on the insights expressed in Subchapter 2.1 Contributions from North America and Australia. The third example presents a new case study that explores the use of the ^{137}Cs resampling technique on a third experimental watershed (EFW) of the same experimental area of *Crotone*. It was not yet published, and it will be presented here in full detail. In this third study some of the assumptions made in the previous contributions will be tested and discussed and some preliminary conclusions on the application of the resampling approach will be formulated.

The second subchapter (2.2.2 Contribution from Spain) presents a new case study carried in the experimental area of *The Sola*, Northern Spain, where the ^{137}Cs resampling method was used to quantify soil redistribution rates in mountain agroecosystems by the team of Dr. L. Gaspar and Prof. A. Navas from Aula Dei Experimental Station in Zaragoza.

2.2.1 Contributions from Italy: Using a ^{137}Cs Resampling Technique to Investigate Soil Erosion and Sediment Mobilisation in Small Uncultivated Catchments

The first attempt to compare estimates of soil erosion rates obtained using ^{137}Cs measurements with independent long-term measurements of soil erosion was undertaken by Porto et al. (2001) in Southern Italy. Subsequent studies demonstrated the validity of the traditional technique at different spatial scales and for different land uses (Porto et al. 2003; Di Stefano et al. 2005; Porto et al. 2011; Porto and Walling 2012a,b; Porto and Callegari 2022). Several years after these first applications of the ^{137}Cs technique, some of these researchers pioneered the use of the resampling approach by revisiting some of the sites that had been previously sampled for ^{137}Cs measurements (see Fig. 2.24).

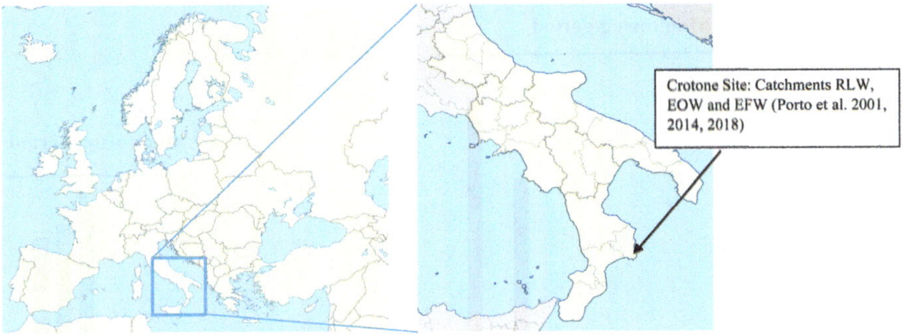

Fig. 2.24 Study sites in Italy that were resampled for ^{137}Cs analysis by Porto et al. (2014, 2018)

The results of these studies will be discussed in the following three subchapters. In the first two subchapters (2.2.1.1 and 2.2.1.2) the studies undertaken by Porto et al. (2014, 2018) in two small watersheds designated as RLW (rangeland watershed) and EOW (eucalyptus open grove watershed) will be reviewed and discussed in the light of new information provided for these watersheds. The third subchapter (2.2.1.3) provides an unpublished data from a new study undertaken in a third watershed designated as EFW (eucalyptus forest watershed).

2.2.1.1 Porto et al. (2014) Approach

The resampling approach reported by Porto et al. (2014) was the first attempt to explore the potential for using repeated soil sampling for ^{137}Cs measurements to provide information on changing soil erosion rates undertaken in Europe. This study focused on a small (1.36 ha) watershed covered by eucalyptus forest (EOW, in original publication designated as W2), located in the experimental area of *Crotone* (Calabria), Southern Italy. This watershed has never been cultivated. In 1968 an eucalyptus forest (*Eucalyptus occidentalis*) was planted, and it was cut twice in 1978 and 1990. Prior to afforestation this watershed was overgrown by rangeland vegetation with no anthropogenic disturbance. In 1990 it was attacked by eucalyptus pest *Phoracantha semipunctata* (a species of beetle originating from Australia), which caused many trees to die. This contributed to loosening the eucalyptus forest and transforming it to grove with open canopy. The fallen trunks of dead trees were not removed from the grove and remained in the EOW. The watershed had been instrumented for rainfall, runoff, and sediment yield measurements since 1978 and these data were available for two different periods. The first period extended from 1978 (the beginning of the monitoring) to 1994, when the measurements ceased due to lack of funds and equipment malfunctioning. The measurements restarted at the end of 2005, when the equipment was refurbished, providing a second period of data collection extending from 2006 to 2012 (Fig. 2.25).

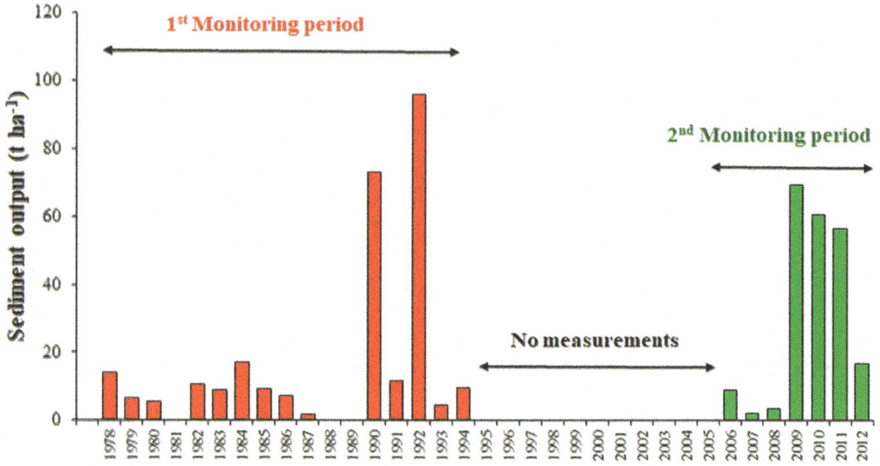

Fig. 2.25 Measurements of annual sediment yield available for EOW

Taking account of the lack of values of annual sediment yield for three of the years during the first period (1981, 1988, 1989), a total of 21 years of sediment yield data were therefore available for the catchment outlet and these data were used to compare the measured sediment yields with the estimates of net soil loss derived from the catchment obtained using the repeated ^{137}Cs measurements.

The first comparison of the sediment yield measurements and the estimates of net soil loss from the catchment obtained using ^{137}Cs measurements was undertaken in 2001, based on a first sampling campaign undertaken during two separate sampling campaigns in 1998 and 1999 (Porto et al. 2001). The sampling involved 55 sampling points located in the studied catchment and an undisturbed area with no evidence of erosion or deposition, which was selected as a reference site in the vicinity of the studied catchment. The ^{137}Cs conversion to soil erosion rates was done using a *profile distribution model (PDM)* in first study (Porto et al. 2001) and a refined *diffusion and migration model (DMM)* was used later (Porto et al. 2003). These two conversion models provided estimates of long-term soil erosion rates that were compared with the measurements of sediment yield provided by the first monitoring period (1978–1994). The mean value obtained by averaging the individual estimates of soil loss produced for the 55 sampling points was compared with the value of mean annual sediment yield available for the catchment outlet. The comparison of annual sediment yields measured at the catchment outlet with the mean annual net soil loss from the catchment obtained by ^{137}Cs method was considered acceptable, since the channel system within the catchment was poorly developed and was seen as providing neither a significant additional source of sediment unaccounted for by the ^{137}Cs measurements nor a significant store or sink for sediment mobilized from the catchment slopes. The estimate of net soil loss from the catchment derived from the ^{137}Cs measurements using the PDM model was in a range of 28.4 (t ha^{-1} yr^{-1}) to 29.5 (t ha^{-1} yr^{-1}) and was about 40% greater than the mean annual sediment yield of 20.8 (t ha^{-1} yr^{-1}) for the 1978–1994 period measured at the catchment outlet. This latter value was calculated using the 14 years of measurements available for this 17-year period. In contrary, the estimated net soil loss obtained from the DMM (Porto et al. 2003) due to its more comprehensive physical basis was much lower, ranging from 11.8 (t ha^{-1} yr^{-1}) to 14.7 (t ha^{-1} yr^{-1}) depending on the shape of the ^{137}Cs reference profiles used to calibrate the model. One reason of this significant difference was the different time period represented by the two data sets and great temporal variability. The annual sediment yields in this environment are characterized by considerable inter-annual variability and therefore a high coefficient of variation. Consequently, the soil erosion rates measured during the different time windows , i.e. the 1954–1998/9 period (represented by the mean annual net soil loss measured by ^{137}Cs method) and the 1978–1994 period (represented by the sediment yield at the catchment outlet) are significantly different.

The EOW catchment was resampled in 2013 and the resampling followed the original sampling design, and 55 replicate samples were taken at points adjacent to the 55 sampling points used in the original sampling campaign. No reference samples were collected as the ^{137}Cs reference inventory for the resampling data set was calculated using the value obtained in 1998 (2604 Bq m^{-2}) decay corrected to 2013 (1849 Bq m^{-2}). The 55 new estimates of soil loss obtained using the DMM were used to calculate the net soil loss from the catchment that could be compared with the mean annual sediment yield at the catchment outlet. For this second comparison, the sediment yield measurements extended to 21 years of monitoring (1978–1994 and 2006–2012) and provided a more representative value for the mean annual sediment yield (28.4 t ha^{-1} yr^{-1}). The net soil loss obtained for the second sampling campaign (13.6 t ha^{-1} yr^{-1}) using DMM, which relates to the 1954–2012 period, was very similar to that (14.0 t ha^{-1} yr^{-1}) derived from the first campaign and corresponding to the same reference profile. The analysis suggested that no recognisable change in the rate of net soil loss from the catchment could be detected over the entire period considered. This conclusion was refined by applying the traditional reference approach (TRA) to estimate the net soil loss from the catchment for the period between the sampling campaigns (i.e. 1999–2012). This involved the following steps:

1. Estimating the mean annual net soil loss (t ha^{-1} yr^{-1}) from the catchment for the 1998–1999 sampling campaigns was used to calculate the total soil loss (t ha^{-1}) for the 1954–1998 period, assuming that the time window associated with the ^{137}Cs measurements started in 1954 with the commencement of ^{137}Cs fallout.
2. Similarly, the mean annual soil loss (t ha^{-1} yr^{-1}) obtained for the 2013 sampling campaign was used to calculate the total soil loss (t ha^{-1}) for the 1954–2012 period.
3. The difference between these two values of total soil loss provides an estimate of the total soil loss for the 1999–2012 period.

The value of total soil loss of the 1999–2012 period was used to estimate the mean annual net soil loss (t ha^{-1} yr^{-1}) of this period. The results of the above calculations provided a mean net soil loss of 12.6 (t ha^{-1} yr^{-1}) from the catchment of the 1999–2012 period. Comparison of this value with that of the 1954–1998 period (14.0

t ha^{-1} yr^{-1}) confirmed that no significant change in erosion rate occurred during the time window since the first sampling campaign until the resampling. To draw meaningful conclusions from these results, the uncertainty associated with the above estimates was also considered (Porto et al. 2014). The results indicated that there was no clear difference between the mean annual erosion rate of the 1954–1998 period and that of the 1999–2013 period, suggesting that no significant difference in the rate of net soil loss from the study catchment could be detected between the two examined periods.

Discussion: A key feature of the study reported by Porto et al. (2014) is the satisfactory agreement between the estimates of net soil loss from the study catchment obtained using the TRA with the independent measurements of sediment yield available for the investigated catchment. From this perspective, the results obtained by this study are essentially unique, because the data available for the EOW catchment monitored by the University Mediterranea were, at the time of the study, the only example of long-term measurements of sediment yield from a small catchment available in Italy for the time window covered by ^{137}Cs measurements (since 1953). Another useful outcome of this study is showing that the time elapsed between the two sampling campaigns (13 years) was sufficiently long for soil erosion to cause well detectable reduction of the ^{137}Cs inventories as it was documented by the two sampling campaigns. However, if we plot these inventories against the 1:1 line for comparison (see Fig. 2.26), some of the 1998 inventories (decay-corrected to 2013) are of a similar magnitude to, or less than, the corresponding values related to the sampling undertaken in 2013. More particularly, the Fig. 2.26 indicates that for 20 of the 55 sampling sites, the ^{137}Cs inventory measured in 2013 was greater than that obtained in 1998, if the latter values are decay corrected to 2013. This suggests that deposition could have occurred at these 20 sites during the period between the two sampling campaigns.

If we apply the TRA for each sampling point these potential deposition sites can be identified (see Fig. 2.27).

The authors accounted for this inventory gain by the fact that the points sampled in 2013 were not exactly the same as those sampled in 1998 and argued that the inventories for the individual sampling points measured by the two sampling campaigns

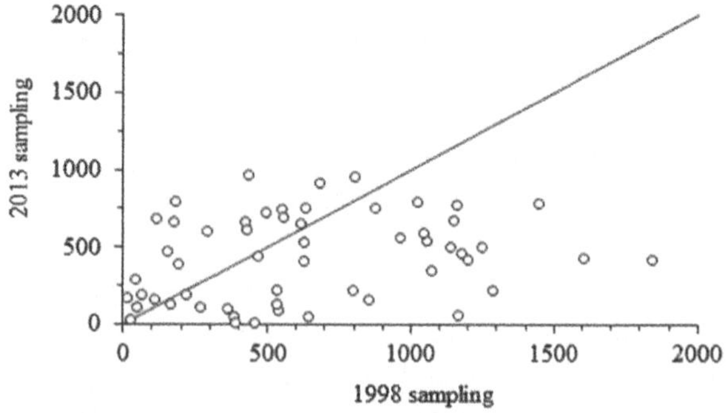

Fig. 2.26 Comparison of the ^{137}Cs inventories (Bq m^{-2}) measured at the 55 sampling points in EOW in 1998 and 2013; The inventories measured in 1998 have been decay corrected to 2013 (based on Porto et al. 2014)

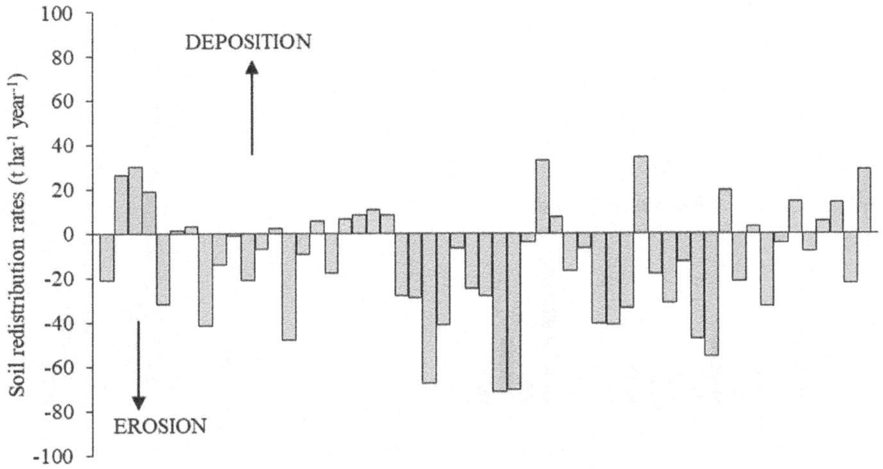

Fig. 2.27 Comparison of the soil redistribution rates estimated for the corresponding sampling sites of the sampling campaigns in 1998 and 2013; The plotting positions for the sampling points along the horizontal axis simply represents sample number (i.e. 1–55)

could not be directly compared because of the natural small-scale variability of the ^{137}Cs inventories and the spatial heterogeneity of soil redistribution rates associated with the uncultivated soils of the studied catchment. For this reason, the two sets of samples had been treated as independent datasets and the TRA was applied to the two estimates of mean annual erosion rate obtained by averaging the two sets of 55 single values.

2.2.1.2 Porto et al. (2018) Approach

Another study using the ^{137}Cs resampling approach was reported by Porto et al. (2018). It was carried in watershed overgrown by rangeland (RLW, in original publication designed as W1) of similar size (1.47 ha) as EOW. The land use of this watershed was very stable, and the rangeland vegetation occur here without any changes during the last 100 years. RLW was firstly sampled for ^{137}Cs in 2001 (Porto et al. 2004) and resampled in 2014. Like EOW, RLW also has been instrumented to monitor rainfall, runoff, and sediment yield since 1978 and here the gained data are available for two different periods. The first period lasted since 1978 (the beginning of the monitoring) until 1994, when the measurements ceased due to lack of funds and equipment malfunctioning. The measurements restarted at the end of 2008, when the equipment was refurbished, and a second monitoring period lasted since 2009 until 2015 (Fig. 2.28).

In total 23 years' time series of sediment yield data is available for the catchment outlet, providing another opportunity to compare the longer-term mean annual value

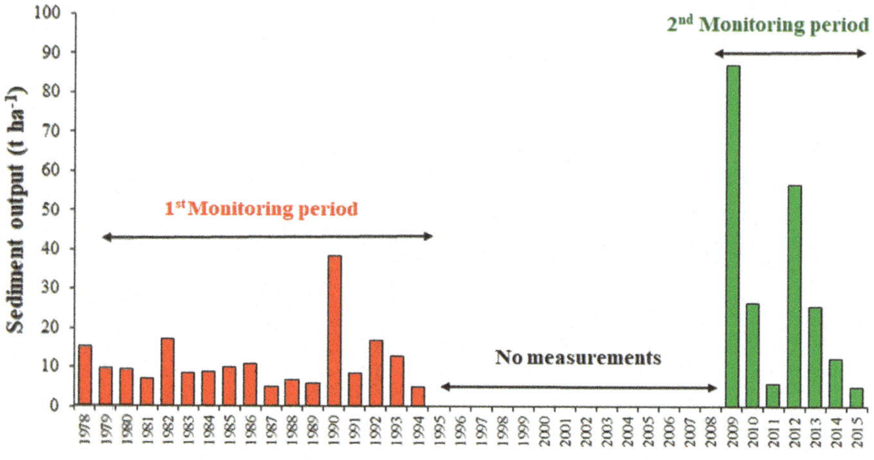

Fig. 2.28 Sediment yield measured at RLW

provided by these measurements with the estimates of net soil loss obtained using the ^{137}Cs technique.

The first ^{137}Cs sampling was undertaken in 2001 (Porto et al. 2004) and 68 sampling points were sampled. For RLW the same reference value was used as for EOW. The refined DMM was used for calculating the erosion rates (Porto et al. 2004). The resulting estimates of long-term mean annual soil loss were compared with the sediment yield measurements at the catchment outlet carried during the first monitoring period (1978–1994). The mean erosion rate of the 68 sampling points, which provided an estimate of the mean annual net soil loss from the catchment, was compared with the mean annual sediment yield measured at the catchment outlet. The mean annual net soil loss calculated from the ^{137}Cs measurements was slightly lower than the mean sediment yield at the catchment outlet (Porto et al. 2004) and some depositional areas were identified in the watershed. The DMM model provided an estimate of mean annual net soil loss of 8.6 t ha^{-1} yr^{-1}, whilst the sediment yield measurements for the 1978–1994 period indicated a mean annual sediment yield of 11.6 t ha^{-1} yr^{-1}. Also in this study, the uncertainties involved in comparing a mean annual sediment yield measured at catchment outlet (representing 17 years of measurements) with the ^{137}Cs-based estimate (covering ca. 55 years period) resulted in the difference between the two compared values.

The RLW was resampled in 2014 and 68 replicate samples were collected from points close to the points sampled in the first campaign. A new ^{137}Cs reference value was obtained using new samples collected from an undisturbed area close to that used for reference sampling during the first sampling campaign. The new reference value (1796 Bq m^{-2}) corresponded very closely to that obtained in 1998 decay corrected to 2014 (1806 Bq m^{-2}). The 68 new estimates of soil redistribution calculated by the DMM were averaged for comparison with the mean annual sediment yield at the catchment outlet

. For this second comparison, the sediment yield measurements extended to 23 years of monitoring (1978–1994 and 2009–2014) and provided a more representative value for the mean annual sediment yield. The net soil loss from the catchment for the 1954–2014 period obtained by the ^{137}Cs method (9.4 t ha^{-1} yr^{-1}) was slightly higher than the value (8.6 t ha^{-1} yr^{-1}) resulting from the first campaign, but lower than the long-term sediment yield measured at the catchment outlet (17.8 t ha^{-1} yr^{-1}). The uncertainties involved in comparing an estimate of mean annual sediment yield based on 23 years of monitoring with the ^{137}Cs-derived estimate of net soil loss which relates to a period of ca. 60 years, must, however, be recognized.

Discussion: Porto et al. (2018) did not attempt to calculate the erosion rate for the period between the two sampling campaigns (2002–2014). Recently the TRA was applied for RLW as for EOW. The results provided a mean value of net soil loss of 12.3 (t ha^{-1} yr^{-1}) and confirmed that the DMM, coupled with the TRA, shows a tendency to underestimate the erosion rates as compared to the sediment yield measurements at the catchment outlet (35.9 t ha^{-1} yr^{-1}) available for the second monitoring period (2009–2014).

A key feature of the study undertaken by Porto et al. (2018) is that, with EOW, this study permitted a comparison of net soil loss from the watershed provided by the ^{137}Cs technique with long-term measurements of sediment yield provided by monitoring at the catchment outlet. However, these measurements were not available for the entire period covered by the ^{137}Cs technique (1954–2014) and the uncertainty associated with them must be considered. This and the other sources of uncertainty will be an objective of further discussion in the next subchapter, which deals with a third catchment study. In addition, it is important to recognize that the direct comparison of the net soil loss from the catchment slopes with the measured sediment yield at the catchment outlet assumes that there is no long-term storage of sediment in the channel system of the catchment and that the channel does not provide an additional sediment that would be included in the measured sediment yield, causing the measured sediment yield to exceed the net soil loss from the slopes.

2.2.1.3 Example of Using the ^{137}Cs Resampling Technique to Investigate Recent Changes in Soil Erosion Rates in a Small Catchment in Southern Italy

Introduction

Validating the ^{137}Cs resampling method requires two conditions. Firstly, the sites where independent information on soil redistribution rates available for comparison with ^{137}Cs derived erosion rates are needed. Existing long-term measurements of soil loss that cover a period sufficient to permit a meaningful comparison with results obtained by ^{137}Cs method are limited to only a few sites around the world. Secondly, to provide a contemporary test of the effectiveness of the ^{137}Cs resampling method, these sites would require ^{137}Cs measurements carried out long time ago to provide the baseline for current resampling (i.e. the time between the first

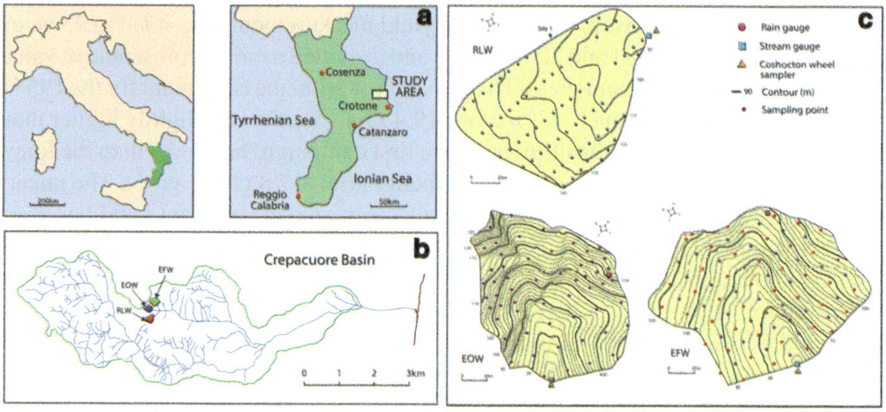

Fig. 2.29 Location of Crotone Experimental Area with the three experimental watersheds (RLW, EOW and EFW) and positions of rain and stream gauges and ^{137}Cs sampling points

and second ^{137}Cs sampling should be sufficiently long for soil redistribution to be evidently manifested). This requirement further reduces the availability of validation sites to a limited number of locations. To date, the only existing studies that address fully these requirements known to the authors of this guidelines were carried in Canada (Kachanoski 1987) using data from ten experimental plots, located in Southern Ontario, with 11 years of soil loss measurements from 1966 to 1976 (subchapter 2.1), and the field site named '20NB' located near the town of Grand Falls in New Brunswick, for which 15 years of measurements, from 1990 to 2005, are available (see Tiessen et al. 2009 and Subchapter 2.1). However, both these sites are cultivated, and tillage operations are conducted repeatedly over the years. To date, the only examples of existing study sites with uncultivated soils (and that concerns both requirements necessary for the comparison), are available in the *Crotone* Experimental Area with three study watersheds (RLW, EOW, EFW), located in Calabria, Southern Italy.

Study Area

The results obtained for RLW and EOW have been summarized in the previous two chapters. The third study was conducted in the third watershed of the Crotone Experimental Area (Fig. 2.29) having a similar size (1.65 ha) as EOW and RLW and being occupied by eucalyptus forest (EFW). It was reforested by eucalyptus in 1968 similarly as EOW but this forest has been cut three times in 1986, 2006 and 2016.

This third study provides an overview of the results of erosion measurements in EFW and their comparison with the values measured in EOW and RLW. EFW was instrumented like EOW and RLW in 1978, to monitor the effects of afforestation on its hydrological response and sediment yield (Porto et al. 2009). Long-term records of rainfall, runoff, and sediment yield, albeit intermittent due to periodic malfunctioning of the sampling devices, are available for the period 1978–2015 to provide the estimates of net soil loss required to validate both the traditional single sampling and the resampling ^{137}Cs technique, assuming the two requirements discussed in Subchapters 2.2.1.1 and 2.2.1.2 (lack of sedimentation in channel and riparian zone and lack of channel and gully erosion) are met.

Methods

Assessing Soil Redistribution by Suspended Sediment Measurements and ^{137}Cs Technique

The soil redistribution in RLW, EOW and EFW was expressed by suspended sediment yield measurements at the watershed outlets. The technical details of the watershed outlet instrumentation and procedure of suspended sediment measurements is provided by Porto et al. (2014, 2018).

It was assumed that the sediment yield measured at the watershed outlet is equivalent to the net soil loss from the watershed slopes if no storage of sediment in the stream channel or riparian zone, neither the channel erosion exists. If there would exist a storage of sediment originating from the slopes in the channel or riparian zone, the measured sediment yield would underestimate the net soil loss. On the other hand, if some portion of sediment measured at the watershed outlet is coming from channel erosion or gully erosion, the measured sediment yield would overestimate the net soil loss from the slopes estimated using ^{137}Cs method.

Furthermore, the ^{137}Cs measurements are only able to provide an estimate of sheet and rill erosion but not gully erosion, which is mobilizing also deeper subsoil material free of ^{137}Cs contamination. It is therefore possible that the estimate of net soil loss provided by ^{137}Cs measurements would underestimate the net soil loss from the catchment slopes. Direct comparison of net soil loss provided by ^{137}Cs measurements with sediment yield measurements at the watershed outlet has to assume that all sediment mobilized from the catchment is eroded by surface processes (sheet and rill erosion) and does not involve deeper subsoil (i.e. gully erosion).

Soil Sampling for ^{137}Cs Determination

The investigation of erosion rates in the watershed using the ^{137}Cs technique involved two sampling campaigns undertaken in 1999 and 2001 and in 2014–2015 (a resampling interval of ca 15 years). The sampling procedure adopted to collect soil samples within the catchment was essentially the same for both campaigns. In each case, two replicate cores were collected from each of the 83 sampling points and subsequently bulked to obtain a single sample from each sampling point for ^{137}Cs measurements. The positions of the sampling points had been originally selected using a grid supplemented by additional samples positioned to reflect the local topography and they were seen to provide representative coverage of the catchment.

During the resampling campaign in 2014–2015, it was not possible to locate the sampling points used in the first campaign precisely. However, the samples were collected within a maximum of 2 m from points of previous sampling, avoiding the tree trunks and the spots disturbed by the first sampling. The reference site, located in the vicinity of the studied watershed, was originally sampled in 1999. In that campaign, one depth incremental soil profile was sampled using a scraper plate and six depth incremental samples were collected from the same area using a soil corer (diameter of 8.6 cm) to take account of the spatial variability of ^{137}Cs activity (Porto et al. 2001). These cores were sectioned to 2 cm depth increments and their total inventories and ^{137}Cs depth distributions were compared with that provided by the scraper plate. Considering the larger sampling area (652 cm^2), the inventories and the associated profile distributions obtained with the scraper plate were used as the reference values.

In 2014, the reference site sampled in 1999 was resampled to obtain a new ^{137}Cs reference value. To reduce the measurement uncertainty associated with the reference inventory, the new inventories were obtained after collection of 8 soil samples from a square area (25 m^2 in size) using the same coring device as previously but equipped with a larger core tube (diameter of 11 cm). Each individual sample was sectioned into 1–4 cm depth increments. All the corresponding depth incremental subsamples from 8 cores corresponding to one depth increment were combined and a single bulked sample for each depth increment was obtained and analyzed.

Sample Preparation and ^{137}Cs Analyses

All samples collected during the sampling campaigns were initially oven dried at 105 °C, disaggregated, homogenized and passed through a 2 mm sieve. A representative fraction of the < 2 mm fraction of each sample was then placed into a 0.5 L Marinelli beakers or, in the case of smaller samples, to plastic pots (330 cm^3), for determination of their ^{137}Cs activity. The samples collected during the first campaign in 1999–2001 were analyzed at the gamma spectrometry laboratory of the Department of Geography in the University of Exeter, UK using high resolution HPGe detector. The samples collected in 2014 and 2015, during the resampling campaign, were analyzed at the Department of Agraria of the University Mediterranea of Reggio Calabria, Italy, using equivalent detector. All measurements of ^{137}Cs had a precision of ca ±10% at the 95% level of confidence. The total inventory (Bq m^{-2}) of each bulk core was calculated as the product of the measured ^{137}Cs activity (Bq kg^{-1}) and the dry mass of the < 2 mm fraction of the bulk core (kg), divided by the surface area associated with the core (m^2).

Model for Conversion of ^{137}Cs Inventories to Soil Erosion Rates

To convert the ^{137}Cs measurements into values of soil loss and deposition, a refined version of the diffusion and migration model (DMM) was adopted (Porto et al. 2003). It assumes that the evolution of the ^{137}Cs depth distribution within a soil profile through time is dependent on the atmospheric fallout input to the soil surface

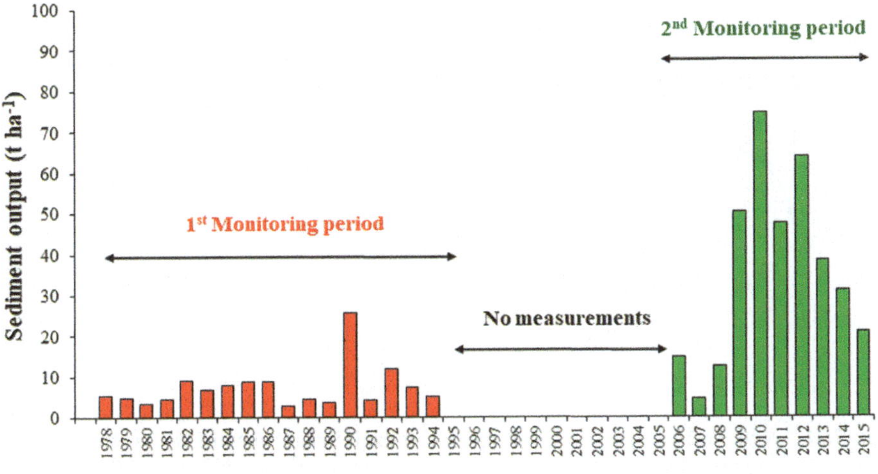

Fig. 2.30 Values of sediment yield measured in EFW

and on its post-depositional redistribution. If we assume a constant soil lowering E (kg m^{-2}) due to erosion and consider the redistribution processes to be essentially diffusional, the activity of ^{137}Cs within the soil profile can be reproduced by the following equation:

$$C(x,t,t') = e^{-\lambda_{Cs}(t-t')} \int_0^\infty \frac{I(t')}{H} e^{-\frac{y}{H}} \left\{ \left[e^{-\frac{[(x+E)+y]^2}{4D_{Cs}(t-t')}} e^{-\frac{[(x+E)-y]^2}{4D_{Cs}(t-t')}} \right] \frac{1}{\sqrt{4\pi D_{Cs}(t-t')}} \right\} dy \tag{2.12}$$

where:

C(x,t,t')	is the ^{137}Cs activity related to the mass depth *x* and time *t'* (Bq kg^{-1});
DCs	is the time-averaged estimate of the diffusion coefficient ($kg^2 m^{-4} yr^{-1}$);
H	is the relaxation mass depth i.e. the ^{137}Cs initial distribution along the soil profile following the atmospheric fallout (kg m^{-2});
λ_{Cs}	is the constant of radioactive decay for ^{137}Cs (0.023 yr^{-1});
x	is the soil mass depth measured from the soil surface downwards (kg m^{-2});
t	is the time elapsed since the first fallout (yr);
I(t')	is the ^{137}Cs atmospheric flux at time *t'* (Bq $m^{-2} yr^{-1}$).

Results

Suspended Sediment

The sediment yields measured at the catchment outlet are presented at Fig. 2.30.

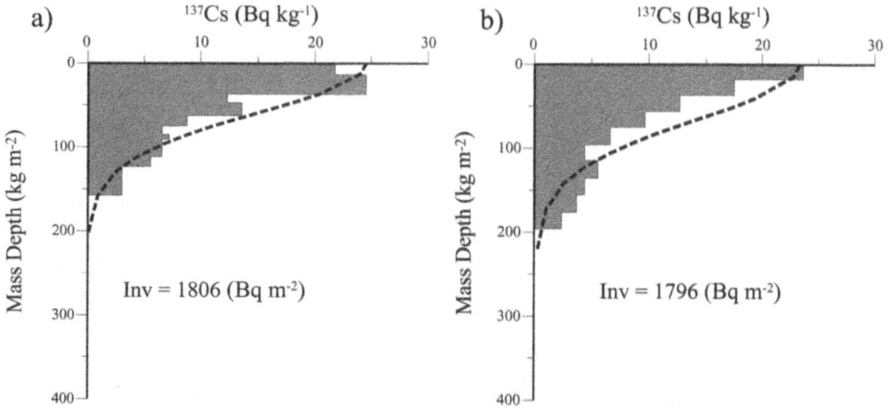

Fig. 2.31 Depth distribution of ^{137}Cs inventories of the reference site in 1999 (**a**) and 2014 (**b**); The dashed lines represent the depth distributions modelled using the DMM model (Eq. 2.11)

In total 27-year time series is available and these data represent two separated periods of 1978–1994 and 2006–2015. During the second period the sediment yields were much greater than during the first one.

Radiocaesium Distribution

The depth distribution of ^{137}Cs at the reference site for both the 1999 (decay corrected to 2014) and 2014 sampling campaigns (Fig. 2.31) shows a steady exponential decline of ^{137}Cs concentration with depth which is a typical feature of undisturbed sites where the original fallout ^{137}Cs inventory was influenced just by radioactive decay and shallow diffusion caused by bioturbation of soil mesofauna. This depth distribution is a basic input for conversion models for calculating soil erosion at non-cultivated land (pastures and forests).

The profiles reported in the above figure show similar ^{137}Cs activities and depth distributions. The total inventory associated with the profile sampled in 1999 (decay corrected to 2014) is estimated to 1,806 Bq m^{-2}. For the 2014 profile, the total inventory is estimated to 1,796 Bq m^{-2}, so both values are in good agreement. The ^{137}Cs depth distributions predicted for 1999 and 2014 by the *diffusion and migration model (DMM)* are also reported (see dashed lines on Fig. 2.31). This model proved to be very effective in the experimental sites located in southern Italy, as reported by Porto et al. (2003, 2016b) and Porto and Callegari (2022) where more details can be found.

The ability of DMM to simulate the depth distribution of ^{137}Cs activity in soil profiles and to estimate soil erosion and deposition rates has been documented elsewhere (Porto et al. 2003).

Information on the mean values, ranges and standard deviations of the ^{137}Cs inventories of RLW, EOW, EFW is provided in Table 2.1. These values of first sampling campaigns have been decay corrected to the date of the resampling campaign.

Table 2.1 Overview of ^{137}Cs inventories for the reference site and the studied watersheds (values from the 1998–2001 sampling campaigns decay corrected to time of corresponding resampling campaign)

Watershed	1998–2001				2013–2015			
	Reference inventory	Watershed inventories			Reference inventory	Watershed inventories		
		Mean	Range	SD		Mean	Range	SD
	(Bq m^{-2})	(Bq m^{-2})	(Bq m^{-2})	(Bq m^{-2})	(Bq m^{-2})	(Bq m^{-2})	(Bq m^{-2})	(Bq m^{-2})
RLW	1806	1035	3–3,008	565	1796	768	21–3,127	592
EOW		642	13–1,844	446		435	6–969	280
EFW		802	71–1,795	501		462	14–1,354	290

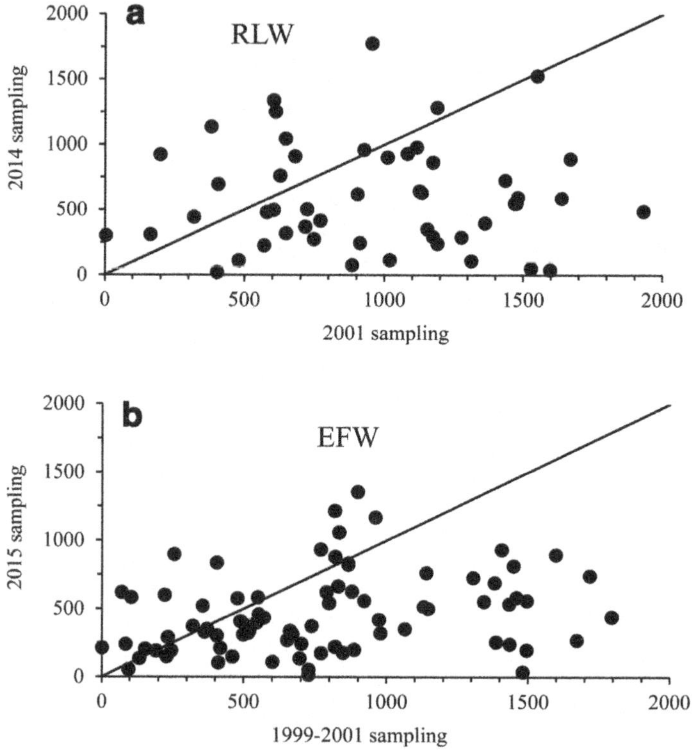

Fig. 2.32 Comparison of the ^{137}Cs inventories (Bq m^{-2}) of the same sampling points in 2001 (decay corrected to 2014–15) and 2014–15 in RLW (a) and EFW (b) watersheds

Table 2.1 indicates that the ^{137}Cs inventories of EOW and EFW (except for only 1 sampling point) are lower than the reference inventory, indicating that both watersheds have been affected by soil loss over the period covered by the ^{137}Cs measurements. In the case of RLW, the inventories for some cores exceeded the reference value, indicating depositional areas. However, if the ^{137}Cs inventories of RLW and EFW measured during both sampling campaigns are plotted one against the other (see Fig. 2.31a,b), it is clear that some of the inventories from the first sampling campaigns (decay corrected to the date of the resampling campaigns) are of the same magnitude or lower than the corresponding values documented by the resampling campaigns.

For RLW (Fig. 2.32a), the existence of some areas of deposition is indicated in Table 2.1 by inventories exceeding the reference inventory. However, in the case of EFW (Fig. 2.32b), where none of the inventories recorded by either sampling campaign exceeded the reference inventory for that campaign (Table 2.1), the plot of the inventories obtained for the individual sampling points for the two sampling campaigns is similar to that presented in Fig. 2.26 for catchment EOW. In EFW, for 19 out of 81 sampling points (i.e. ca. 25% of all sampling points) the inventory in 2015 was greater than in 2001, suggesting that deposition had occurred between the first and the second campaign. This is, again, inconsistent with the other evidence and clearly demonstrates that the inventories obtained for the individual points cannot be directly compared because of great microvariability of ^{137}Cs distribution due to bioturbation and different soil properties. These factors can play important role as the position of sampling points of the first and second campaign cannot be exactly equal. Therefore, the assessment of soil redistribution during the first and second time windows, should be based on mean soil redistribution rate calculated from all sampled points, which is a much more representative value.

Estimates of Soil Redistribution Rates in EFW

The DMM (see Eq. 2.11) was applied separately for each sampling campaign (2001 and 2015) using the traditional approach, whereby each inventory value was compared with the reference value related to the corresponding sampling campaign. The resulting estimates of soil redistribution rates for this watershed, obtained for the period 1954–2000, range between −0.1 t ha^{-1} yr^{-1} and −26.5 t ha^{-1} yr^{-1} (with a mean of −11.3 t ha^{-1} yr^{-1}), while those related to the period 1954–2015 are respectively −3.2 t ha^{-1} yr^{-1} and −26.8 t ha^{-1} yr^{-1} (with a mean of −12.4 t ha^{-1} yr^{-1}). These estimates are reported in Table 2.2 together with the corresponding values of RLW and EOW.

Information reported in Table 2.2 provides a basis for applying the TRA to obtain the soil redistribution rates for the period between the first and the resampling campaigns (i.e. 2001–2015). The calculation procedure described previously for EOW was used.

These estimates are reported in Table 2.3 together with the corresponding values of sediment yield obtained from the watershed outlets.

In this case, the suspended sediment yield estimates representing the same periods as the estimates of net soil loss provided by the ^{137}Cs measurements are also provided. These are based on the measured values at the catchment outlet, with values for the missing years estimated using the rainfall data for those years.

Uncertainty Analysis

Applying the TRA to estimate the mean annual net soil loss for the period between the two sampling campaigns and comparing these values with measured sediment yield data for the same period involve several sources of uncertainty associated with ^{137}Cs measurements and sediment yield datasets. There were identified four main factors contributing to uncertainty:

1. Reliability and precision of the laboratory equipment used for the measurements of ^{137}Cs.
2. Sampling variability associated with the small-scale heterogeneity of ^{137}Cs inventories in undisturbed soils, which affects both the reference value and the individual sampling points at study site.

Table 2.2 Soil redistribution rates derived from ^{137}Cs measurements using the DMM for RLW, EOW and EFW during two time windows (negative values indicating soil erosion)

Watershed	First time window	Soil redistribution rates (t ha^{-1} year^{-1})			Second time window	Soil redistribution rates (t ha^{-1} year^{-1})		
		Mean	Range	SD		Mean	Range	SD
RLW	1954–2001	−8.6	5.0–40.7	6.82	1954–2014	−9.4	10.2–25.7	6.43
EOW	1954–1998	−14.0	0–34.6	7.72	1954–2012	−13.6	− 6.4–30.0	5.63
EFW	1954–2001	−11.3	−0.1–26.5	6.22	1954–2015	−12.4	− 3.2–26.8	4.42

Table 2.3 Comparison of soil redistribution rates derived from ^{137}Cs measurements for the period between the two ^{137}Cs samplings (2001-2015)and the suspended sediment yield at watershed outlet obtained for the second monitoring period (2006–2015)

Watershed	Suspended sediment			Suspended sediment	Erosion rates (^{137}Cs technique)
	Second time window			Period between the two sampling campaigns	
	(t ha^{-1} year^{-1})			(t ha^{-1} year^{-1})	
	Mean	Range	SD	Mean	Mean (TRA)
RLW	31.5	5.1–87.4	30.4	22.4	12.3
EOW	25.3	2.3–69.3	26.1	20.6	12.6
EFW	35.9	4.6–74.7	23.2	29.3	16.1

3. Ability of the conversion model to represent the physical processes involved in ^{137}Cs redistribution and availability of data on model input parameters.
4. Efficiency of the sampling device used to obtain the sediment yield measurements and appropriate interpolation of the sediment yield estimates for the periods when no data are available.

Understanding these factors may help to reduce their impact and minimize the uncertainty of data processing. The identified factors contributing to uncertainties were assessed as follows:

1. The first source of uncertainty is related to the precision of measurements provided by the gamma spectrometry. In our case, all the individual values of ^{137}Cs inventory associated with the soil cores collected from the three watersheds are estimated to have an uncertainty of ca. ±10% at the 95% level of confidence. This reflects the counting time. However, the uncertainty of ca. ±10% is associated with individual sampling points, but the overall uncertainty associated with the mean value of all sampled points can be assumed to be minimal because the ^{137}Cs inventories are underestimated for some point and overestimated for other points and it is reasonable to assume that underestimation at some points would be cancelled out by overestimation at others. The fact that the samples were analyzed in different laboratories, also needs to be considered. Subsamples of all the sets of soil cores collected during the first and resampling campaigns have been retained in the soil sample archive of the University Mediterranea of Reggio Calabria. This made it possible to undertake a comparison between the analyses carried out in the different laboratories involved in this study. In this respect, some of the samples analyzed at Palermo in 1998, Exeter in 1999–2001 and Basel in 2012 have been reanalyzed at Reggio Calabria and decay corrected to the date of sampling to explore their accuracy. In Fig. 2.33 this comparison is illustrated for three sets of samples that have been used in this study. More particularly, the new measurements (expressed in Bq m^{-2}) obtained in Reggio Calabria were compared with the measurements in other labs:

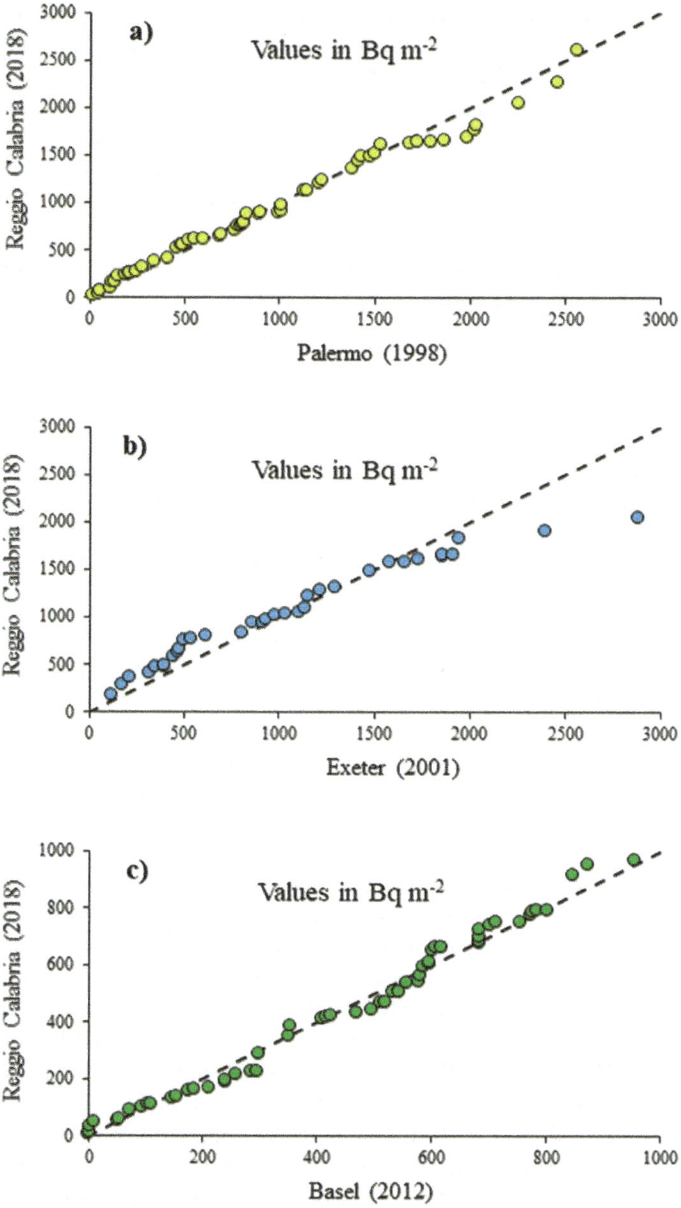

Fig. 2.33 Comparison between the results of the ^{137}Cs analyses carried out in Palermo, Exeter and Basel and analyses carried in Reggio Calabria: a) EOW sample set (55 points) analyzed in Palermo in 1998; b) EFW sample set (34 points) analyzed in Exeter in 2001; c) EOW sample set (55 points) analyzed in Basel in 2012

The results presented in Fig. 2.33 indicate that, for each comparison, the agreement between the two corresponding sets of measurements is acceptable and that any bias associated with the use of a specific detector is limited. To confirm this conclusion, a Mann-Whitney test was used to test the significance of the difference between the corresponding samples. The resulting P-values are in all cases lower than the critical value at the 0.05 level of significance, indicating that there is no significant difference between the two sets of samples for each comparison. However, based on the values of P obtained in this test, a value of ±11% at the 95% level of confidence has been attributed to this source of uncertainty, when comparing the estimates of mean annual net soil loss from a catchment provided for different periods by the ^{137}Cs measurements.

2. The second source of uncertainty reflects the small scale ^{137}Cs variability associated with a sample collected from a single sampling point. Such sampling variability is likely to be greater for uncultivated soils, such as those sampled in the studies reported here, than for cultivated soils where the topsoil is regularly homogenized by tillage. Studies conducted by Sutherland (1996) suggested that this uncertainty can be as high as ±30% at the 95% level of confidence. Obviously, this uncertainty will be important when considering the estimates of soil redistribution rate derived from a single point, but when the sampling strategy is based on a large number of samples and the study area is relatively small (in this study, 136, 110, and 162 sampling points are considered for catchments RLW, EOW and EFW, respectively, over an area of ca. 1.5 ha) the underestimation at some points should be compensated by the overestimation at others and the overall uncertainty will be limited and can reasonably be ignored. The same assumptions can be made regarding the uncertainty associated with the reference value. In this case, a sampling strategy involving collection of several sectioned cores from an undisturbed site made it possible to take the small-scale spatial variability of the ^{137}Cs reference inventory into account and avoid introducing uncertainty into the estimation of the reference inventory. Nevertheless, uncertainty associated with the sampling design and more importantly sample size and sample number must be incorporated into any uncertainty analysis. The estimates of mean annual soil loss associated with the two sets of samples will have a standard error, the magnitude of which will reflect the number of sampling points. Based on the standard deviations associated with the mean annual soil loss for the two studied periods, a mean standard error of ±13% at the 95% level of confidence has been attributed to the mean annual soil loss (see Porto et al. 2014 for details).
3. The third source of uncertainty can be ascribed to the conversion model and its calibration parameters. The choice of the DMM was made on the basis of it being a physically based model which takes into account the key physical processes involved. Use of the DMM may generate some bias in the estimates of soil redistribution provided for each individual sampling point, but when the comparison involves the mean soil redistribution rates for two periods the uncertainty associated with the relative magnitude and direction of any change is likely to be limited. However, when the estimates derived from ^{137}Cs measurements

are compared with independent measurements of sediment yield for the same period this uncertainty should be considered. In the absence of detailed guidance for quantifying this source of uncertainty, a nominal ±25% has been used.

4. The precision of the sampling device used to measure the sediment yield at the catchment outlets was estimated by Porto et al. (2014) to ca. ±15% at the 95% level of confidence. However, if the aim of the study is to compare the mean annual net soil loss from the watershed obtained by ^{137}Cs method for the different time windows with sediment yields measured at the watershed outlet, there is a problem with different periods reflected by sediment yield measurements and the ^{137}Cs measurements. The ^{137}Cs measurements cover a period extending back to 1954, while the periods of sediment yield measurements are usually much shorter. To provide a meaningful comparison for this study, the available measurements of annual sediment yield have been extrapolated to cover the same periods as the estimates of net soil loss provided by the ^{137}Cs measurements using the annual rainfall totals for those years and a relationship between annual sediment yield and annual rainfall for all three individual catchments. An uncertainty of ca ±50% at the 95% level of confidence has been attributed to the values of mean annual sediment yield when these involve extrapolated data.

The uncertainty caused by Factors 1, 2, and 3 can be combined using the root mean square error propagation method of Topping (1972), known as summation in quadrature, and shown, for our study site, in the following equation.

$$Unc_{ne} = \sqrt{11^2 + 13^2 + 25^2} \tag{2.13}$$

The estimate of uncertainty caused by Factors 1, 2, and 3 associated with the mean annual net soil loss generated by the ^{137}Cs measurements provided by Eq. (2.13) gives an overall uncertainty of ±30% at the 95% level of confidence. An equation of the same form has been applied to combine the two estimates of uncertainty caused by Factor 4, 15% and 50% respectively. In this case, an estimate of uncertainty of ±52% at the 95% level of confidence was obtained for the values of mean annual sediment yield associated with each catchment.

Discussion

The collected data (Tables 2.2 and 2.3) were used to compare the mean annual net soil loss from the three studied watersheds estimated by the ^{137}Cs method for the two different time windows, namely the period since the commencement of ^{137}Cs fallout (1954) until the first sampling and the period between the two sampling campaigns and further, these values were compared with the sediment yield at the watershed outlet for these two time periods for each of the three watersheds (Fig. 2.34).

Based on the results obtained by the uncertainty analysis, the estimates of mean annual net soil loss and sediment yield have been expressed as ranges and related to the results obtained using the TRA. When comparing the mean annual net soil

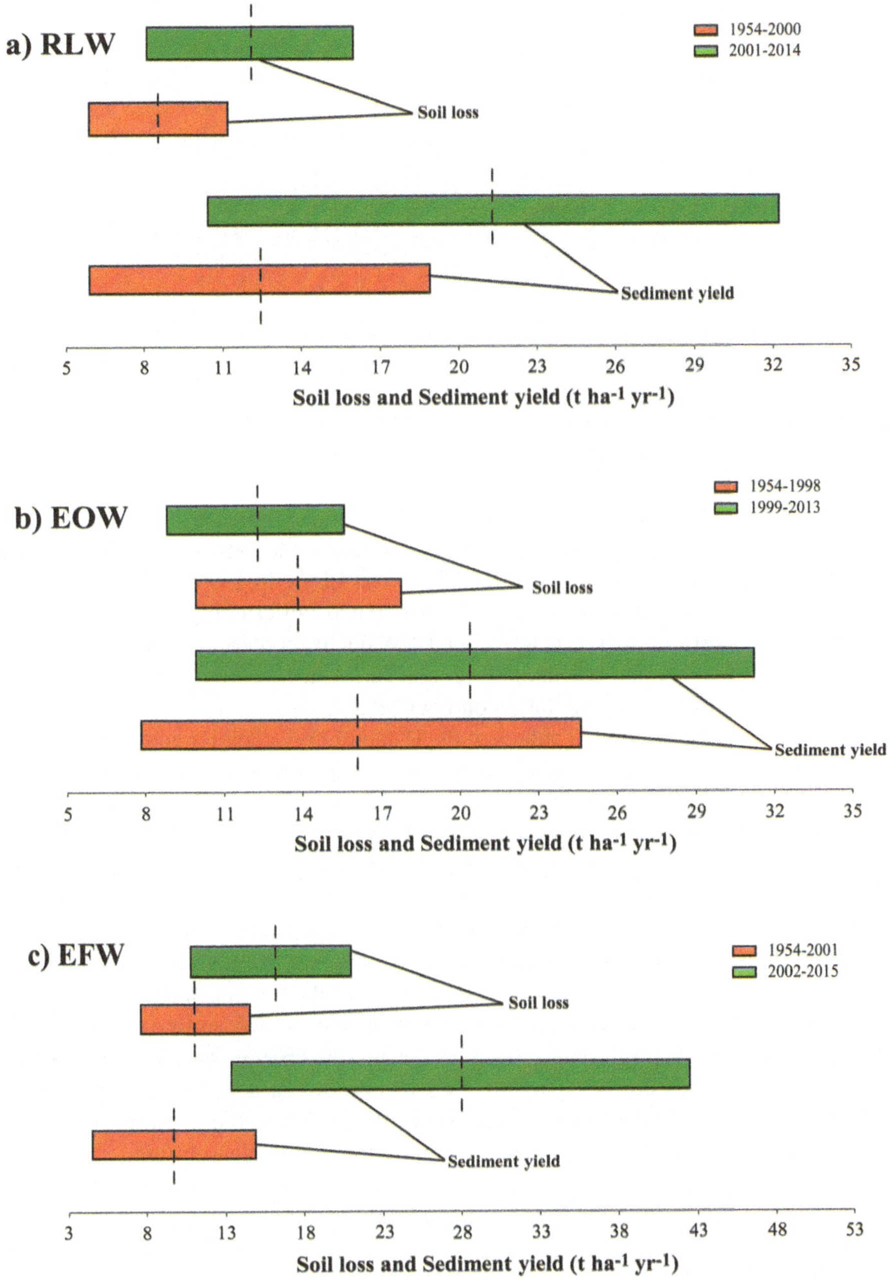

Fig. 2.34 Comparison of the ranges (confidence limits) associated with the mean annual net soil loss from watersheds (^{137}Cs derived data) and sediment yield at watershed outlets for each study catchment for the periods related to the sampling campaigns, considering their uncertainty

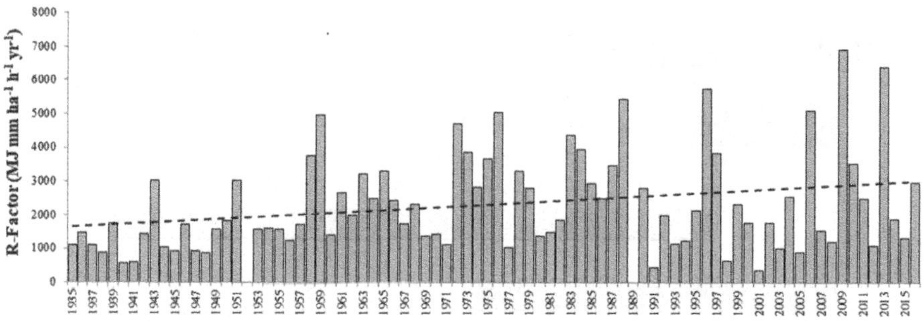

Fig. 2.35 Trend of annual values of rainfall erosivity for the study area

loss provided by the ^{137}Cs measurements with the mean annual sediment yield at the watershed outlet for the same periods, Fig. 2.34 suggests that the two datasets are not significantly different.

However, comparing the two investigated periods, both the mean annual net soil loss estimated by ^{137}Cs method and sediment yield at the watershed outlet showed higher values for the second period in RLW and EFW. In EOW this trend is not so pronounced. The sediment yield at the watershed outlet is higher in the second investigated period similarly as in RLW and EFW but its increase is not so pronounced as in RLW and EFW, and the annual net soil loss estimated by ^{137}Cs method has even decreased slightly during the second period.

Figure 2.35 presents annual values of the R-factor (Wischmeier and Smith 1978), calculated using the 30-min rainfall records of some representative stations located in the area. The estimates of annual erosivity shown in Fig. 2.35 provide evidence of an increasing trend that is particularly prominent for the last 10 years, because of the occurrence of an increased number of very severe rainfall events during that period. The estimates of net soil loss provided by the ^{137}Cs measurements are, as stated above, in agreement with this trend of increasing rainfall erosivity in RLW and EFW. This trend is weaker in EOW, what could result from some environmental factors other than rainfall, which occurred only in EOW but not in RLW and EFW.

A first possible explanation could be the influence of clearcutting that occurred in 1978 and 1990. It might cause an increase of soil erosion rate in some part of the watershed expressed by changed ^{137}Cs inventories at several sampling points sampled during the first sampling campaign (1998). Thus, the erosion during the first investigated period could be encouraged by this forest clearcutting.

A second reason could be the occurrence of an eucalyptus pest in this watershed. After clearcutting in 1990 this forest was gradually restored by natural regrowth, but it was attacked by *Phoracantha semipunctata* (a species of beetle originating from Australia) and the tree mortality resulting from this attack caused several trees to fall down and these were not removed from the catchment. The dead timber created sediment traps and may have caused some sampling points to experience deposition rather than continuing erosion. Even if the ^{137}Cs inventories were all lower than the reference value, temporary deposition may have occurred in recent years, and this may have reduced the mean erosion rate during the second investigated period.

In view of the above results and considering the associated uncertainty of the analyses, it is important to recognize that even if the estimates of mean annual net erosion rates provided by the ^{137}Cs measurements are lower than the corresponding measured value of sediment yield, the results presented above showed a clear evidence of a general increase of soil erosion rates in the second investigated period and it was demonstrated that the ^{137}Cs resampling method can be considered a useful tool to investigate the development trends of erosion processes.

Conclusions

The three studies described above have demonstrated both the successful application of the ^{137}Cs resampling approach in small watersheds in southern Italy and its potential for more general application to document changing erosion rates. They have also highlighted a number of issues that merit further thought or investigation. These are outlined below.

1. All three studies were undertaken in uncultivated watersheds. In general, uncultivated land is characterized by lower erosion rates than cultivated, but in these study areas the erosion rates were relatively high. Both the relatively high erosion rates and the exponential depth distribution of ^{137}Cs found in their uncultivated soils mean that the ^{137}Cs inventories were sensitive to erosion and thus conditions were favorable for the application of the ^{137}Cs resampling approach. Significant changes of inventories occurred between the two sampling campaigns. Further work is, however, required to explore and define the limiting conditions for successful use of the resampling technique, in terms of erosion rates and the duration of the interval between sampling campaigns. This must take into consideration the uncertainty associated with measurements of ^{137}Cs inventories. The time window between the two sampling campaigns must be sufficiently long for erosion to cause the change of ^{137}Cs inventory significantly greater than what is the uncertainty associated with measurements of ^{137}Cs inventories. These limiting conditions will be different on cultivated land, where the mixing of ^{137}Cs over the plough layer would make the ^{137}Cs inventory less sensitive to erosion and deposition than in areas with uncultivated land.
2. The presented studies have focused on the use of ^{137}Cs resampling. Further attention should be directed to the relative merits and potential for using unsupported Lead-210 ($^{210}Pb_{ex}$) resampling (see Porto et al. 2016a). This fallout radionuclide is characterized by a very different fallout input (see Porto et al. 2006).The ^{137}Cs fallout was time limited and primarily occurred in a relatively short period between the mid-1950s and the early 1970s. Since that time the soil redistribution and radioactive decay have caused changes in the inventory. The situation is different for $^{210}Pb_{ex}$ where fallout can be viewed as effectively constant through time and it replenishes the depletion of the inventory caused by erosion

and radioactive decay. This characteristic, coupled with the shorter half-life of $^{210}Pb_{ex}$ means that this radionuclide may provide a better basis for the resampling technique, since it will be more sensitive to recent changes in soil redistribution rates.

3. In the studies reported above, it was assumed that the ^{137}Cs inventories at the individual sampling points of the two sampling campaigns are related to the same sampling points. However, this is not true because of both practical and theoretical constraints concerning the sample positions. It was practically not possible to locate the sampling points during the resampling campaign precisely at the same sampling points as during the first sampling campaign because recording the sample positions during the first sampling campaign were not sufficiently precise. But, even if they would be sufficiently precise, theoretically it is not possible to sample the same point, because the soil at the sampling point was disturbed by the sampling so that this place does not represent anymore the representative undisturbed soil. It might be suggested that this problem can be solved by collecting a soil core situated as closely as possible to original sampling point, but not at the point disturbed by the former sampling. However, it is well known that uncultivated soils are characterized by significant small-scale heterogeneity of the ^{137}Cs inventory, as demonstrated by the relatively high coefficient of variation of soil cores collected from a small area, and this approach was therefore seen as being problematic. In the case of cultivated soils, the mixing of soil by tillage is known to reduce the small-scale heterogeneity. Also, it will greatly reduce the effects of local disturbance in the immediate vicinity of the original sampling point since the soil around the original sampling point will be mixed by tillage operations. It is, nevertheless, still possible that the action of removing a core will have a long-term effect on the inventory at the sampling point, because the core samples usually exceed the depth of ploughed horizon. Further work is needed to demonstrate empirically to which extent can the first and second sampling represent the ^{137}Cs inventory of the "same" sampling point. In the studies reported above, this problem was overcome by sampling adjacent points, rather than the same point. The two sets of samples collected by the two sampling campaigns were treated as independent sample sets that provided representative information on the ^{137}Cs inventories and therefore longer-term soil redistribution rates in the study catchment on the two occasions. This made it possible to estimate the mean annual rate of net soil loss indicated for the periods between the onset of ^{137}Cs fallout and the time of the first sampling and the period between the two sampling campaigns.
4. A preliminary attempt has been made to identify and quantify the uncertainty associated with the estimates of mean annual net soil loss provided for the study catchments by the ^{137}Cs measurements. Further work is needed to develop this aspect of the study further. Such estimates of uncertainty are very important when attempting to demonstrate changes in erosion rates in response to changes in land use or land management or climate change. Differences in erosion rates demonstrated by ^{137}Cs measurements using the resampling technique must be

greater than the uncertainty associated with those estimates if they are to be meaningful.

5. The reported studies are unique in providing independent estimates of sediment yields and information on recent changes in those sediment yields for the individual catchments which could be compared with the estimates of net soil loss provided by the ^{137}Cs measurements. This is useful for assessing the changing erosion rates in the study catchments. Perhaps more importantly, the close agreement between the two measures of sediment export from the catchments provide clear evidence of the ability of ^{137}Cs method to provide meaningful and reliable estimates of soil redistribution rates. This is important in view of the recent criticism by Parsons and Foster (2011). It is, however, important to consider the uncertainties associated with both estimates. In addition, it is necessary to recognize that for such direct comparisons between estimates of net soil loss based on ^{137}Cs measurement and measured sediment yields a number of conditions should be met if the two estimates should be closely similar. In general terms, such comparisons are only likely to be valid for small catchments. These conditions include:

    a. An assumption is that estimates of net soil loss provided by ^{137}Cs measurements represent the amount of sediment delivered from the catchment slopes to the channel system. If these estimates are to be compared with the sediment yield at the catchment outlet, there should be no evidence of sediment deposition below the foot slopes, in the channel system or adjacent riparian zone during its transfer to the catchment outlet. Such deposition would reduce the amount of sediment reaching the catchment outlet and the measured sediment yield would underestimate the net soil loss.
    b. Further, the channel system should not be a sediment source contributing by additional sediment to that represented by the net soil loss from the slopes. If the channel system would be a significant sediment source, the measured sediment yield at the watershed outlet would be increased and it would overestimate the net soil loss from the slopes.
    c. Estimates of sediment mobilized from the slopes of a watershed derived from ^{137}Cs measurements can reflect only sediment mobilized by surface processes i.e. sheet and rill erosion. They reduce the ^{137}Cs inventory of affected slope as compared to the reference inventory. Other sediment sources, such as gully erosion, tunnel erosion and landslides cannot be detected by ^{137}Cs measurements, since they mobilize deeper part of the soil profile, which is not contaminated by ^{137}Cs.

Further work should focus on exploring the potential differences between estimates of net soil loss and measurements of sediment yield under a range of scenarios linked to those discussed above.

2.2.2 Comparing Soil Redistribution Estimates of Two Selected Time Windows Using the ^{137}Cs Resampling Approach in Mountain Agroecosystems of Northern Spain

2.2.2.1 Introduction

The fallout radionuclide ^{137}Cs has been successfully used to document rates of soil erosion in a wide variety of landscapes worldwide (Walling and He 1999; Ritchie and McCarty 2008; Kato et al. 2010; Mabit et al. 2013), including Mediterranean environments (Navas et al. 2005; Porto and Walling 2012a, b; Benmansour et al. 2013; Gaspar et al. 2013a; Lizaga et al. 2018). Since the beginning of the 1990s, the EEAD gamma lab pioneered in Spain the analysis of FRNs to trace soil redistribution and evaluated soil loss at different spatial scales in a large variety of environments, from arid and semiarid (Navas and Walling 1992; Quine et al. 1994) to subhumid and temperate (Gaspar and Navas 2013) including various land covers and land use changes (Navas et al. 2017). The records obtained during the past decades are of value to serve as the basis to assess if any changes related to external factors have influenced the mobilization of soil. This can be done if some sites sampled already in past are investigated again using the ^{137}Cs resampling approach.

A great advantage of the ^{137}Cs technique known as resampling or repeat-sampling approach has been employed to document changes in rates of soil redistribution, including both erosion and deposition. This approach is done by subdividing the period between the occurrence of the ^{137}Cs fallout and the time of the resampling campaigns (Li et al. 2011; Porto et al. 2014, 2018). In recent years it has been pointed out that the ^{137}Cs resampling can be an efficient strategy to distinguish and apportion the impact of climate variability and agricultural management on soil loss or gain. Another important advantage of ^{137}Cs resampling is that it also allows evaluating the effectiveness of soil conservation measures (Schuller et al. 2004; Moustakim et al. 2019).

Using this approach requires fulfilling two basic conditions:

1. The research is performed in an area already investigated with FRNs in the past;
2. The lapse of time between the two FRN sampling campaigns is sufficiently long to notice significant changes of soil redistribution magnitudes (at least one decade after the first investigation).

There are some other conditions that should be modified according to the study objectives. If the main purpose should be to investigate the impact of climate change, it would be necessary to select a study site where the land use has remained unchanged since the initial FRN investigation. If the objectives focus on the impact of different land uses, a site should be chosen where the land use has changed, and if the primary objective is to assess the impact of soil conservation, samples should be taken from a study site where the same land use initially occurred without conservation measures, and later, conservation measures were introduced.

In this study, we are testing and validating the ^{137}Cs resampling approach within an upland agroecosystem in northern Spain. The resampling approach is conducted to explore the potential of utilizing repeated soil sampling for ^{137}Cs measurements, aiming to provide insights into evolving soil redistribution rates over a time span between an initial sampling campaign and a subsequent one carried out 13 years later. The chosen framework for this approach encompasses representative Mediterranean mountainous landscapes where the land cover and land use have remained unchanged since the ^{137}Cs fallout. Along a hillslope, we have selected two distinct transects where typical land covers and land uses correspond to forested and arable land, and no alterations have taken place. The obtained results will facilitate a deeper comprehension of changes in the soil redistribution process and the opportunity to assess the impact of agricultural management and climate variability on soil erosion.

2.2.2.2 Methodology

Study Site

The study was carried out at the Sola Site, an upland agroecosystem located in the central part of the Pre-Pyrenees (Northeast Spain) (Fig. 2.36). The climate is subhumid, with two rainy seasons, one in spring and the other in autumn when the cultivated land is bare, and the risk of soil erosion is high. The mean annual precipitation is 595 mm and the mean temperature is 12 °C. To address the resampling methodology, we selected a south-facing hillslope with a valuable ^{137}Cs and soil dataset assembled in the early years of the twenty-first century, in which a groundbreaking research using FRNs was developed in the frame of Spanish National research projects that have documented processes and main factors involved in the soil mobilization (Gaspar et al. 2013a, b, 2019, 2021; Navas et al. 2014).

At the studied hillside, two units with differing angles of slope, levels of vegetation cover, and soil management were chosen. The upper section of the slope is composed of limestone (Muschelkalk facies), which typically forms the more elevated landforms with steep slopes (average slope of 19%) in this region, and it is covered by forest. The lower section of the slope is situated on marls and claystone formations with low permeability (Keuper facies), presenting gentle slopes (average slope of 4.5%), and it is occupied by cultivated land. The soil types are influenced by the geological substrate and the variations in elevation. The steep forested areas are characterized by Leptosols and Calcisols, whereas Regosols and Gypsisols predominantly occupy the smoother lower slopes where cultivation takes place.

Soil Sampling and Laboratory Analysis

During the first sampling campaign carried out in 2003, a total of 11 soil profiles were collected, with 6 profiles taken from forested land and 5 profiles from arable land. Thirteen years later (2016), a second campaign was conducted, revisiting the

Fig. 2.36 Location of the study site and collection of soil samples; The automatic core driller is used to retrieve soil profiles sectioned at 5 cm depth intervals for gamma analysis in the laboratory

same 11 locations, and collecting samples once again. The new sampling points were slightly shifted from the original positions to prevent duplication of points sampled during the initial campaign.

Each site was resampled using the same sampler to ensure consistency with the previous sampling program using an 8 cm diameter automatic steel core driller. The tube was driven automatically into the ground to an average depth of 40 cm, or until the bedrock was reached, and was subsequently extracted manually and sectioned in the field at 5 cm depth increments (Fig. 2.36). All samples were oven dried, disaggregated and sieved (<2 mm). The coarse fraction above 2 mm was separated from the fine fraction under 2 mm that was used for ^{137}Cs analysis.

The ^{137}Cs activities were measured using high resolution, low background, low energy, coaxial hyperpure germanium detector coupled to an amplifier and multichannel analyzer. The detector had an efficiency of 50% and a 1.9 keV resolution and was calibrated using standard certified samples in the same geometry as the measured samples. The ^{137}Cs gamma emission at 661.6 keV was counted for 43,200 s, yielding the results with analytical precision of 3% at the 95% level of confidence. The detection limit for ^{137}Cs was 0.2 Bq kg^{-1}. The ^{137}Cs content in the soil samples was expressed as mass activity (Bq kg^{-1}) and areal inventory (Bq m^{-2}), which was calculated using the soil bulk density (weight of the < 2 mm fraction of the soil sample divided by the sample volume).

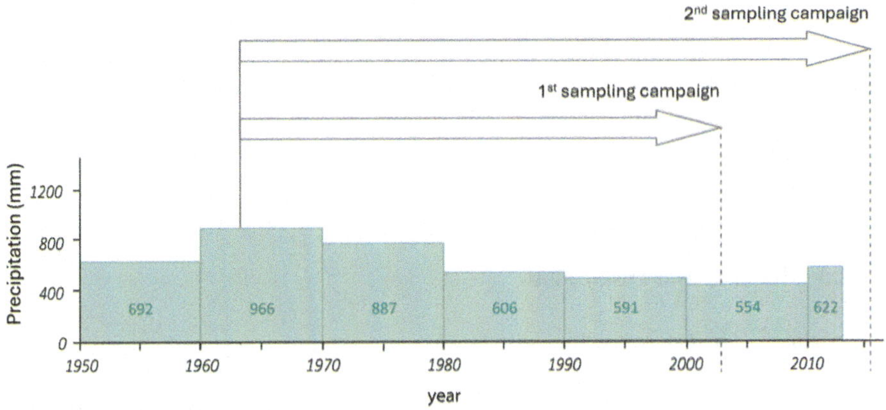

Fig. 2.37 Temporal analysis of rainfall at the study site (10-year means for decades of 1950-2010 period and 3-year mean for 2010-2013)

Climate, Land Cover and Land Uses

Regarding climate, a precise temporal assessment based on daily precipitation has been conducted from 1950 to 2012 using SPREAD, a high-resolution (5 km) daily gridded precipitation dataset derived from raw data collected by 12,858 observatories covering the entire peninsular Spain, as proposed by Serrano-Notivoli et al. (2017). Data beyond 2012 have not yet been published. Figure 2.37 shows the mean annual precipitation every 10 years for our study site. Furthermore, information of an extraordinary rainfall and flood event in a location within the southern Pyrenees in 2012 was recorded (e.g. Gaspar et al. 2019). This event had a regional impact, covering thousands of square kilometers, occurring over three days (19th–21st October), and resulted in an accumulation of approximately 250 mm of precipitation triggering flooding.

The land use has remained the same for both, forest and arable land, over the 13-year period between the two sampling campaigns. The agricultural practices did not undergo any significant change during the last two decades, as interviews with the farmers indicated. Besides of that, the land use history of the study site was assessed using various aerial photos and remote sensing products (such as Landsat, https://idearagon.aragon.es, https://centrodedescargas.cnig.es).

Estimation of the Soil Redistribution Rates

To quantify the soil redistribution rates derived from the ^{137}Cs inventories we used the models developed by Soto and Navas (2004) for uncultivated soils, and by Soto and Navas (2008) for cultivated soils. These conversion models enable to estimate erosion and deposition rates also under specific conditions of soils with high stone content. Stony soils are very common in Mediterranean environments and in mountain landscapes the stone content often ranges between 30 and 50% and in some cases, it can even reach up to 70–80%. In the conversion models the increases or decreases in the content level due to ^{137}Cs fluxes are corrected by the effective soil volume by using the working volume for ^{137}Cs adsorption, which is the one occupied by the fine soil fraction.

2.2.2.3 Results and Discussion

Estimating ^{137}Cs Reference Inventories

The reference inventory for the second sampling campaign was obtained from the decay of the reference inventory for the study area obtained in the first campaign following the equations below.

$$A_T = A_0 * e^{-\lambda T} \tag{2.14}$$

$$\lambda = \frac{0.693}{T_{1/2}} \tag{2.15}$$

where:

A_T is the activity after time T (yr);
A_0 is the initial activity;
λ is the decay constant (ln(2)/T) which is related to the half-life ($T_{1/2}$) of ^{137}Cs (30.23 years).

The ^{137}Cs reference inventory for the first campaign was 1570 Bq m^{-2} and was established from nine cores collected at stable sites with flat surface that were not affected by erosion or deposition (Navas et al. 2013; Gaspar et al. 2013a), while the reference inventory for the second campaign was 1166 Bq m^{-2}.

Profile distribution of ^{137}Cs Inventories

The detailed information on the depth distribution of the ^{137}Cs mass activity was provided by a total of 22 depth incremental soil profiles. By each sampling campaign 11 sectioned soil profiles (6 profiles under forest and 5 profiles on arable land) were collected. The total ^{137}Cs inventory of each profile was compared with the reference inventory in the study site for each campaign. This comparison enabled the identification of points where soil loss, soil gain, or variations within a stable range occurred.

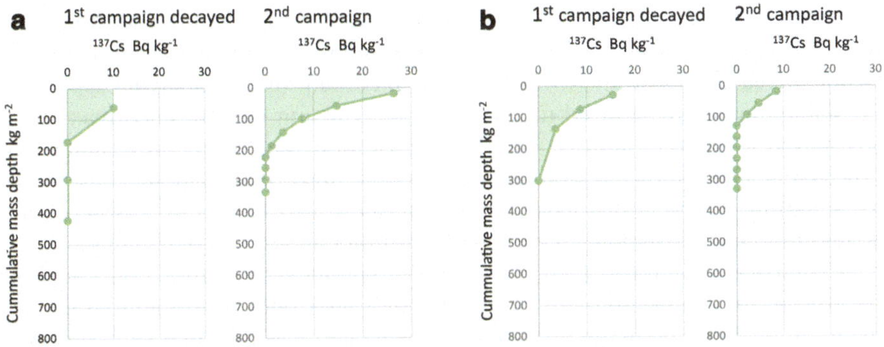

Fig. 2.38 Sectioned soil profiles in forest land

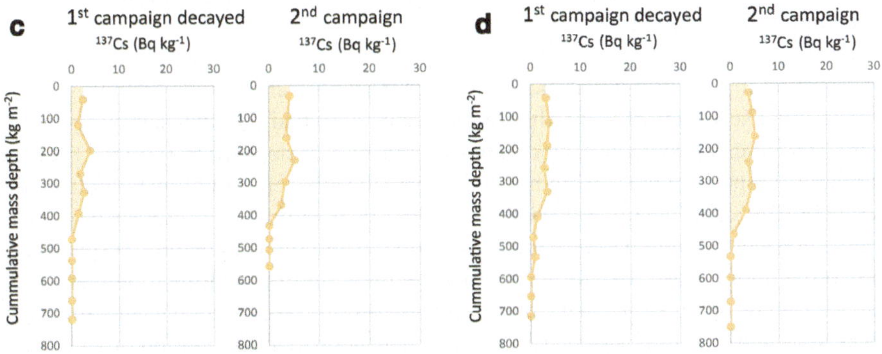

Fig. 2.39 Sectioned soil profiles in arable land

Despite of the ^{137}Cs small-scale variability, the effort of precisely relocating the original sampling points for the second sampling campaign allowed to identify the soil loss and soil gain occurring at each soil profile during the study period. Figure 2.38 shows an example of the changes recorded at two sampling points located on the upper slope under forest. The Profile a experienced deposition with an increase of the inventory by 73%. In contrary, the Profile b experienced erosion with a decrease of inventory by 62%.

Similarly, Fig. 2.39 shows another two examples of sectioned profiles located at the lower slope on arable land. In both cases (Profiles c and d) the inventory increased by 54% and 35%, respectively, suggesting soil deposition.

Soil Redistribution Rates and Sediment Budgets

In the forest the soil redistribution rates varied from low to moderate. In the first campaign, stability was predominant in more than half of the soil profiles. Only one profile recorded erosion while two profiles showed deposition and three were stable. However, in the resampling campaign the number of soil profiles experiencing erosion increased to 4 profiles, whereas the number of profiles displaying deposition and stable conditions reduced to just one each. Erosion rates reached maxima of 0.37 mm yr^{-1} and 0.69 mm yr^{-1} in the first sampling and resampling campaigns, respectively. Deposition rates varied more over the study time window. Thus, for the first sampling campaign the mean deposition rate was 0.16 mm yr^{-1}, whereas for the resampling campaign the maximum deposition rate reached 0.62 mm yr^{-1}.

In comparison to forest land, we observed significantly higher values of both soil erosion and deposition rates in arable land, with rates nearly tenfold greater. The mean erosion rate in the first sampling campaign was 2.70 mm yr^{-1}; however, during the resampling, only deposition was found, leading to a significant increase in soil deposition rates. Therefore, both the minimum and maximum deposition rates were slightly higher in the resampling campaign than in the first sampling, reaching a maximum deposition rate of 3.10 mm yr^{-1}.

Based on previous research reported by Gaspar et al. (2013b) a perceptual model was applied in the study hillslope for the upper and lower sections assuming a 1 m wide strip. The values of soil redistribution rates obtained for each sampling profile were used to calculate equivalent values of soil loss or deposition (kg yr^{-1}) for the individual slope segments, extending halfway to the adjacent sampling point from the neighbouring sampling point in each direction. The resulting values for each segment (Fig. 2.40) were summed to provide a total erosion and total deposition, respectively, thus obtaining net soil loss for each slope segment and the sediment delivery ratio (SDR). For the upper slope, in the first campaign the total soil erosion was estimated at 16 kg yr^{-1} and total deposition at 35 kg yr^{-1}. The net soil accumulation of 19 kg yr^{-1} and a negative SDR (%) indicate that soil redistribution processes were restricted in forest land. However, by the resampling, a greater total soil erosion of 58 kg yr^{-1} was estimated along with a reduced total deposition rate of 26 kg yr^{-1}. The net soil loss of 32 kg yr^{-1} and a SDR of 54% suggest a higher intensity of soil redistribution. For the lower slope, during the first sampling campaign, the total erosion was estimated at 551 kg yr^{-1} and the total deposition at 208 kg yr^{-1}, resulting in a net soil loss of 343 kg yr^{-1} (SDR of 62%). Conversely by the resampling, a net soil accumulation of 441 kg yr^{-1} was obtained.

The sediment budgets assessed for each section confirm that in forest on the upper slope the soil protection by the vegetation cover along with higher soil infiltration rates (Gaspar et al. 2020) limit the soil mobilization. In contrary, on the lower slope with arable land, the soil remains bare after harvest and it becomes more exposed to erosive autumn rainfall that renders soil more vulnerable to be eroded.

Comparing the ^{137}Cs estimates of soil redistribution rates obtained for the two time windows, clear changes in soil erosion and deposition patterns were observed, indicating that higher intensity of soil redistribution processes occurred during the second study period. This suggests that soil redistribution processes gained intensity over the past 13 years. The changes were more evident on arable land than on forest land.

Considering that land cover and land use have remained the same within the study period, the higher soil mobilisation recorded in the time window between both

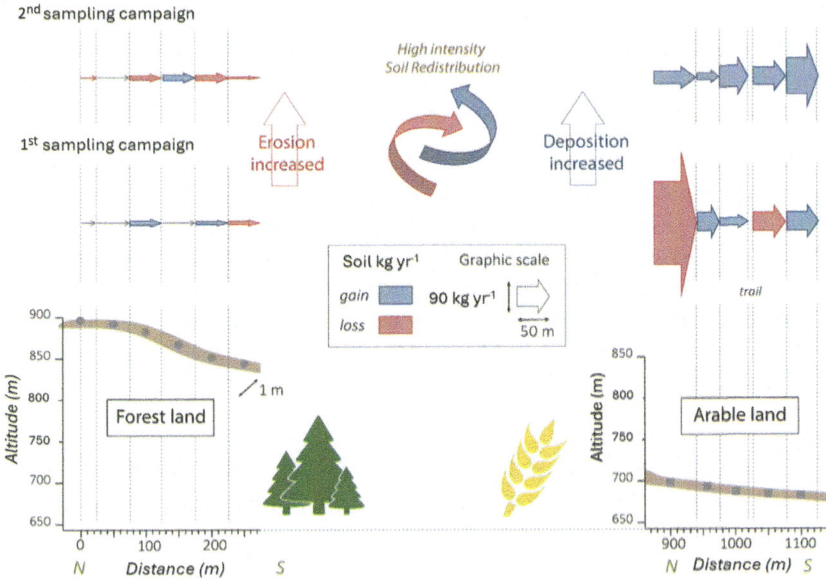

Fig. 2.40 Estimated soil loss and gain (kg yr^{-1}) for a 1 m wide slope segments along the upper slope (forest land) and the lower slope (arable land)

sampling campaigns could be attributed to the increase in precipitation recorded during the last decade. This along with the exceptional rainfall event recorded in 2012 reflect the impact of changes in rainfall patterns.

2.2.2.4 Conclusions

This study is a first attempt to test the ^{137}Cs resampling approach in Spain. The results demonstrated the viability of this technique to assess the temporal dynamics of soil redistribution based on comparing the mean soil erosion and deposition rates for two or more-time windows.

The analyses of the ^{137}Cs inventories recorded during the first sampling and resampling campaign provided detailed information about changes in soil redistribution processes. Specific processes of erosion and deposition were identified from the depth sectioned profiles over the time interval of this study. An increase in the intensity of soil redistribution rates was more evident in arable land than in forest land. Accordingly, a contrasting result in terms of sediment delivery was also found at the upper slope and lower slope.

The obtained results demonstrate the advantage of using depth incremental soil profiles instead of bulk cores for the resampling approach. It allows to verify the coincidence of sampling sites between the first sampling and the resampling, as well as to identify specific soil redistribution processes. The change in redistribution rates recorded in the soil profiles may be associated with a shift in rainfall patterns over the study period because the land management practices have been maintained. The procedure described in this study is a promising application of the ^{137}Cs technique for soil erosion investigations associated with climate change.

Acknowledgments This research was developed within the International Atomic Energy Agency IAEA Coordinated Research Project CRP D1.50.17: Nuclear Techniques for a Better Understanding of the Impact of Climate Change on Soil Erosion in Upland Agro-Ecosystems (Research Agreement No: 20532) and supported by projects PID2019-103946RJ-I00 and PID2019-104857RB-I00 funded by the Spanish Ministry of Science and Innovation (State Research Agency). Thanks are extended to Emil Fulajtar and Lionel Mabit for useful comments.

2.3 Contributions from Asia: Using the ^{137}Cs Resampling Approach to Investigate Soil Erosion Rates in the Loess Plateau, China

The pioneering studies in Asia involving the use of the Caesium-137 (^{137}Cs) in the field of soil erosion and sediment delivery were conducted in China during the late 1980s and 1990s (Zhang et al. 1988; 1989, 1990b). During the following years, in collaboration with some researchers from Europe and USA, Chinese researchers further studied soil erosion process and modern geomorphological evolution (Quine et al. 1997), soil quality-soil redistribution relationship (Li and Lindstrom 2001; Li et al. 2006) using ^{137}Cs method on Chinese Loess Plateau. There are also some studies on wind erosion (Ping et al. 2001; Jiang et al. 2018) and water erosion (Wang et al. 2019) evaluation using ^{137}Cs technique in Qinghai-Tibet and the lower reaches of the Yangtze River. The researchers in many other Asian countries, such as Japan, Iran, South Korea, Turkey and India, conducted many studies to assess soil redistribution rates in different environments using the traditional ^{137}Cs method as well (Takenaka et al. 1998; Karbassi and Amirnezhad 2004; Ruecker et al. 2008; Afshar et al. 2010; Wakiyama et al. 2010; Meusburger et al. 2013; Sankar et al. 2018; Mandal et al. 2019). In addition, Asian countries such as Vietnam, Indonesia, the Philippines, and Thailand have also used the ^{137}Cs method to assess soil erosion in recent years (Suhartini 2006; Anh et al. 2014; Srisuksawad et al. 2015; Olivares et al. 2016). The first study in Asia sampling the ^{137}Cs repeatedly to estimate recent soil erosion rates was conducted in China. Yangjuangou Watershed in the Chinese Loess Plateau was resampled and the new dataset was integrated with already earlier colleced datasets to detect possible change in soil erosion rates for shorter periods (Chappell et al. 2015) (see Fig. 2.41).

The results of these studies undertaken in Yangjuangou (YJG) Watershed will be discussed in the following two subchapters. YJG Watershed selected as the study area is situated in the middle part of Loess Plateau in northern Shaanxi Province, China (36°42′N, 109°31′E). This region is characterized by large topographic variations

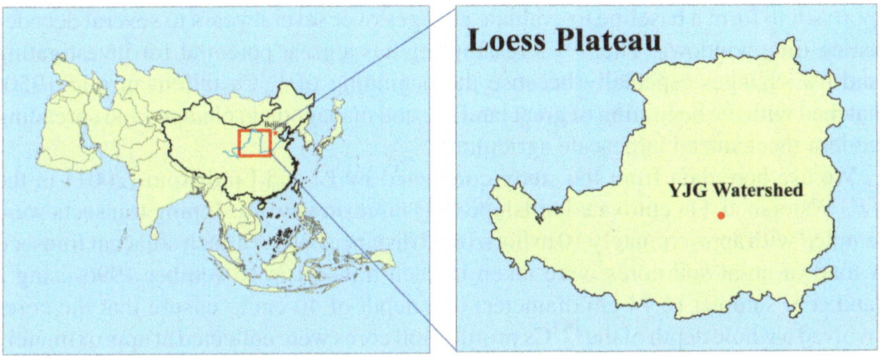

Fig. 2.41 Experimental sites in China that have been resampled for ^{137}Cs analysis by Chappell et al. (2015)

with loess hills and gully land-use types that have maximum altitudinal differences from hilltop to gully bottom of 225 m and a gully density of 2.74 km km^{-2}. The climate is semiarid continental with a mean annual temperature of 13.2 °C, the temperature range from −16.7 to 42.8 °C, and an average annual rainfall of 535 mm that mainly occurs from July to September and exhibits high inter-annual variation. The soil was developed from loess parent material with a uniform soil texture (25% clay, 70% silt, and 5% sand), and is classified as a Calciustepts (Soil Survey Staff 1999) and is very erodible. The soil organic carbon (SOC) concentrations of the plough layer are low (0.4–0.6%) what, together with high silt content is the reason of high erodibility (Li et al. 2006).

Sloping farmland is the main agricultural land in the gully area, and the main crops are sorghum (Sorghum bicolor (L.) Moench), millet (Setaria italica) and beans (Trigonella foenum-graecum). Since 1999, the Revegetation Programme (RP) was implemented at national scale to return sloping farmland to forest and grassland. Currently, the newly introduced vegetation (forest, shrubs or grass) has completely replaced sloping farmlands after 20 years of the RP implementation. Recently, grass, shrubs, forest, and check-dam farmland are the four main land use types that characterize this watershed. The primary grass species are Artemisia Linn, Elymus repens, and small weeds, while the primary shrubs are Prunus armeniaca and Hippophae rhamnoides in mature forests, and Robinia pseudoacacia is the main tree in young forests.

2.3.1 Soil Redistribution Estimated by ^{137}Cs Responding to Land Use Change

The ^{137}Cs technique for estimating net, time-integrated soil redistribution by the processes of wind, water and tillage erosion is increasingly us ing a resampling approach to form a baseline to evaluate changes over several years to several decades lasting time windows. The ^{137}Cs resampling has a great potential for investigating land use changes especially because the beginning of ^{137}Cs fallout in mid 1950s matched with the beginning of great land use and management changes and spreading modern mechanized largescale agriculture.

We use here data from the study conducted by Li and Lindstrom (2001) in the YJG Watershed. On cultivated hillslope (S1) three downward sloping transects were sampled with approximately 10 m horizontal distance between each adjacent transect. A total of eight soil cores were taken in each transect in November 1996 using a hand core sampler (6.74 cm diameter) to a depth of 40 cm to ensure that the cores involved a whole depth of the ^{137}Cs profile. Soil cores were collected at approximately 10 m intervals along the transect. An electronic theodolite was used for precise identification of the sampling point locations. At each sampling point three core samples were collected and bulked to obtain one representative composite sample.

Soil samples for the ^{137}Cs reference inventory were collected on eight undisturbed, non-eroded, uncultivated terraces constructed in 1954 and overgrown by grassland. A mean value of 2390 Bq m^{-2} was determined as the fallout (reference) ^{137}Cs inventory in the study area at the time of investigation.

Following the initiation of the China Grain for Green Program in 1999, the cultivated hillslopes were converted into grassland (Hippophae rhamnoides, Bothriochloa ischaemum, Themeda triandra var. japonica) and forest (Robinia pseudoacacia L.) at the summit, forest on the upper and middle slope and shrubs (Hippophae rhamnoides, Prunus armeniaca L.) on the lower slope and foot slope locations.

In May 2012 (16 years later) soil samples were obtained from approximately the same locations as in the previous study. Global positioning system (GPS) coordinates combined with photographs of the sampling points of the first sampling campaign were used to identify the original sample positions and to make new soil sampling within approximately 2 m from the original sample points. The individual soil sampled from the 24 cores (8 locations along three transects) was retained (not composited) for the measurement of ^{137}Cs.

In both studies, soil samples were air-dried, passed through a 2 mm sieve and weighed. Measurements of ^{137}Cs concentration were conducted on a 1 kg subsample of each bulked core sample using a coaxial hyperpure germaniumdetector coupled to a multichannel analyser. Caesium-137 activity was detected at 662 keV peak with an analytical precision of ±6% requiring counting times of approximately 80,000 s (22.2 h). The results were calculated on a mass specific basis (Bq kg^{-1}) and converted to areal inventories (Bq m^{-2}) using the total weight of bulked core soil and the sampling area.

To convert ^{137}Cs inventories to net soil redistribution (t ha^{-1} yr^{-1}), the Owens (1994) mass-balance model for cultivated soils undergoing erosion and deposition was used. The main parameters used in the calibration models were fixed in the absence of any other information. The tillage depth was set to 0.1 m. The proportion of fallout ^{137}Cs retained in the top 0.5 cm (Q) was estimated based on the ^{137}Cs depth profile measurements to be approximately 0.14 (before decay). The accumulated mass (Ma) of soil to a depth of 0.5 cm (in g cm^{-2}) and that of the tillage layer (Mpl; g cm^{-2})

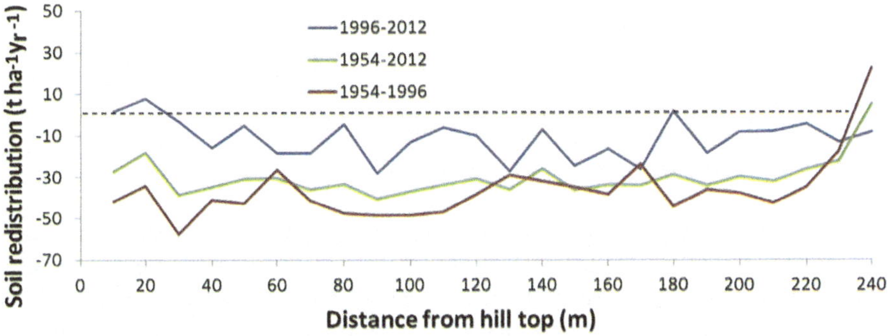

Fig. 2.42 Spatial variability of net soil redistribution at S1 for three time periods; The dashed line is the 0 t ha^{-1} yr^{-1} (i.e. stable land with no erosion and no deposition) and negative values indicate a net loss of soil whilst positive values indicate a net gain of soil

were set to 0.68 g cm^{-2} and 18.6 g cm^{-2}, respectively. Half of the ^{137}Cs deposited at the surface was assumed to be mixed into the soil by tillage ($k = 0.5$). All samples with net ^{137}Cs loss or gain were assumed to have preferential removal (enrichment of ^{137}Cs) of fine material ($P = 1.5$).

The ^{137}Cs reference inventory for 1996 was decay corrected by 16 years to provide values of the reference ^{137}Cs inventory for 2012. These values were used to calculate soil erosion rates from ^{137}Cs inventories of 2012 samples, to estimate net (1954–2012) soil redistribution in the study area using the mass balance model. The resampling approach of the ^{137}Cs technique was applied by decay correcting by 16 years the 1996 sample ^{137}Cs inventories and using them as the reference ^{137}Cs inventory in 2012. These values were used with the 2012 sample ^{137}Cs inventories to estimate net (1996–2012) soil redistribution using the mass balance model.

The results show that within the periods of 1954–1996 and 1954–2012 there is a great variability in soil redistribution along the hillslope (Fig. 2.42).

The spatial soil redistribution patterns along the hillslopes are similar for both time windows, i.e. higher magnitude soil erosion is at the upper slope positions than at medium and lower hillslope positions. There is a detectable reduction over time in net soil redistribution for this hillslope. Net soil erosion was reduced considerably in the recent period of 1996–2012. The reason in this region of China for accelerated soil erosion was increased tillage intensity on steep hillslopes around 1950s. Around the year 2000 cropland was converted to grassland and conservation agriculture (reduced/zero tillage) was adopted in 2002.

2.3.2 Soil Redistribution Changes Assessed Using ^{137}Cs Resampling Under Persisting Land Use

2.3.2.1 Soil Redistribution Change at a Sloping Farmland with 100 Years Persisting Land Use

The ^{137}Cs resampling study was carried out on sloping farmland (S2) which was under the same land use for more than 100 years. Three downward sloping transects were situated within approximately 10 m distance between each adjacent transect. Along each transect eight soil cores were taken in November 1996 using hand core sampler (6.74 cm diameter) to a depth of 40 cm to ensure that the cores invovled the whole depth of the ^{137}Cs contaminated layer. Soil cores were collected at approximately 10 m intervals along the transects. An electronic theodolite was used for precise recording of the sample positions. At each sampling point three core samples were collected and bulked to obtain one representative composite sample.

Soil samples for the ^{137}Cs reference site were taken on eight undisturbed, non-eroded, uncultivated terraces constructed in 1954 and overgrown by grassland. A mean value of 2390 Bq m^{-2} was determined as the fallout (reference) ^{137}Cs inventory for the study area at the time of investigation.

In May 2019, soil samples were taken from approximately the same locations as in the previous study. GPS coordinates combined with photographs of the sampling points of the first sampling campaign were used to identify the original sample positions and to make new soil sampling within approximately 2 m from the original sampling points. The soil samples from the 45 cores were retained (not composited) for the measurement of ^{137}Cs. The laboratory analyses for ^{137}Cs samples are the same as those presented in the Subchapter 2.3.1. The reference ^{137}Cs inventory for 1996 was decay corrected by 23 years to provide values in the study area of the reference ^{137}Cs inventory in 2019. These values were used for interpreting the ^{137}Cs inventories sampled in 2019 to estimate net soil redistribution for the period of 1954–2019 using the Mass Balance II Model (Walling and He 1999). The spatial distribution of net soil redistribution at S2 for two investigated time windows is shown in Fig. 2.43.

The results show that within the periods of 1954–1996 and 1997–2019 there is a great detectable variability along the hillslope. The spatial soil redistribution patterns along the hillslope are slightly variable for each studied time window, but the difference of the mean soil redistribution along the transects between 1954–1996 (-22.9 ± 5.8 t ha^{-1} yr^{-1}) and 1997–2019 (-24.3 ± 4.4 t ha^{-1} yr^{-1}) periods was not statistically significant. The erosion increases along the slope downward. This is because the 140 m long transect has not reached the foot slope. The high variation of the soil redistribution at individual sampling points for the two studied time windows were caused by the high spatial microvariability over the two 2 m distance of soil cores collected during the first and second sampling campaign.

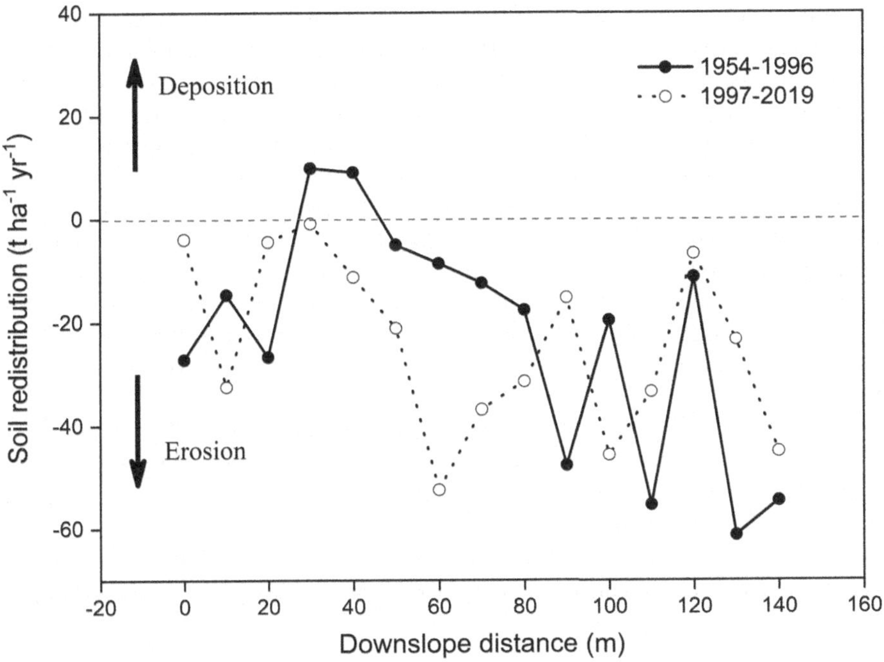

Fig. 2.43 Spatial variability of net soil redistribution at S2 for two time periods (1954–1996 and 1997–2019) in YJG Watershed, Chinese Loess Plateau; The red dashed line is the 0 t ha^{-1} yr^{-1} and negative values indicate a net loss of soil whilst positive values indicate a net gain of soil

2.3.2.2 Changes in Soil Redistribution at Forest Hillslopes with Land Use Persisting Since 1983 and 1965

The ^{137}Cs resampling study was carried out at the two hillslopes with persisting forest but with different ages since the reforestation. Both two artificial forests at S3 and S4 were planted by aerial seeding in 1983 and 1965 to protect cultivated land from land degradation. For each forest hillslope, three downward sloping transects were selected with approximately 10 m horizontal distance between each adjacent transect. A total of eight soil cores were taken in November 1996 along each transect using a hand core sampler (6.74 cm diameter) to a depth of 40 cm to ensure that the cores involve the whole depth of the ^{137}Cs profile. Soil cores were collected at approximately 10 m intervals along the transects. An electronic theodolite was used for precise measurement of the core locations. At each sampling point three core samples were collected and bulked to obtain one representative composite sample.

Soil samples for the ^{137}Cs reference site were taken on eight undisturbed, non-eroded, terraces constructed in 1954 and overgrown by grassland. A mean value of 2390 Bq m^{-2} was determined as the fallout (reference) ^{137}Cs inventory in the study area at the time of investigation.

In June and November 2016, soil samples were taken from approximately the same locations as in the previous study. GPS coordinates combined with photographs of the sampling points of the first sampling campaign were used to identify the original sample positions and to make new soil sampling within approximately 2 m of the original sampling points. The soil samples from the 24 cores (8 locations along three transects) were retained (not composited) for the measurement of ^{137}Cs. The reference ^{137}Cs inventory for 1996 was decay corrected by 20 years to provide values corresponding to the reference ^{137}Cs inventory in 2016. The soil redistribution rates for the period of 1954–2016 were estimated using the Profile Distribution Model (PDM) and the net soil loss was calculated (Porto et al. 2001). The ^{137}Cs inventory from two soil profiles from 1996 and 2016 reference sites was used, respectively, to estimate the shape factors in 1996 ($h_0 = 71$) and 2016 ($h_0 = 18.3$) when using the PDM.

The spatial variability of net soil redistribution along the transects for the two time periods at S3 and S4 forest hillslopes are shown in Figs. 2.44 and 2.45. The results show that within the time periods of 1954–1996 and 1997–2016 there is a great variability at each hillslope.

For S3, the spatial soil redistribution patterns along the hillslope are fluctuating for both time periods, but in general the soil erosion rates are higher in the lower part

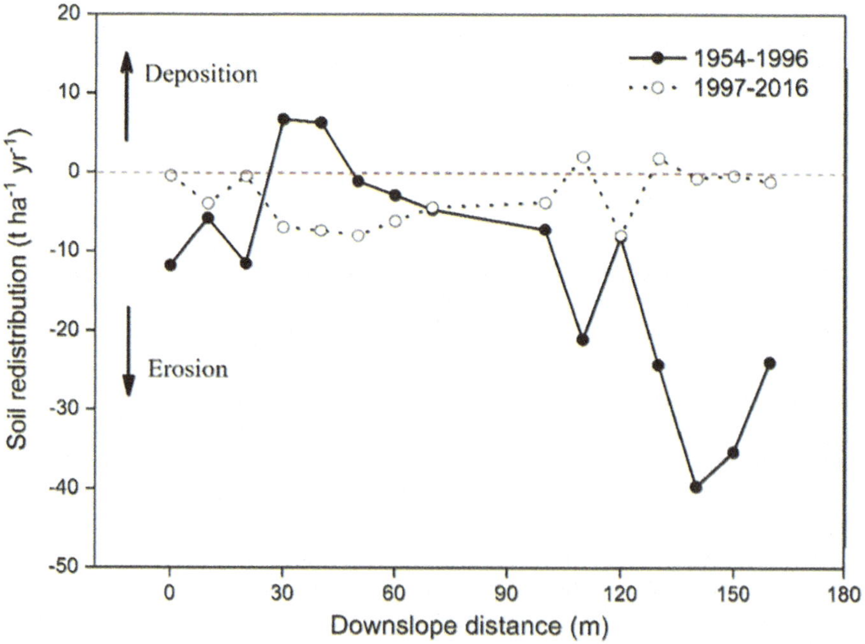

Fig. 2.44 Spatial variability of net soil redistribution along the S3 forest hillslope for two time periods (1954–1996 and 1997–2016) in YJG Watershed; The red dashed line is the 0 t ha^{-1} yr^{-1} and negative values indicate a net loss of soil whilst positive values indicate a net gain of soil

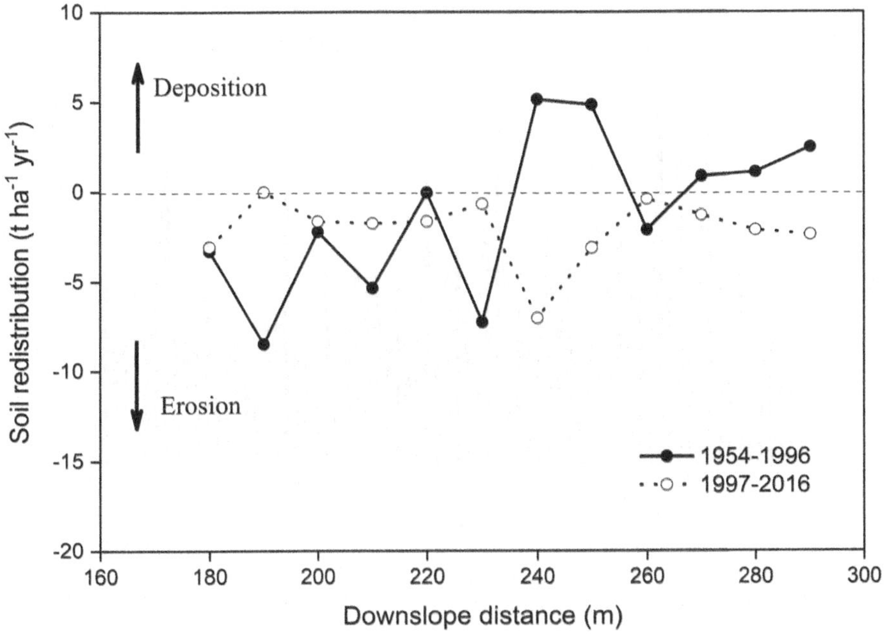

Fig. 2.45 Spatial variability of net soil redistribution along S4 forest hillslope for two time periods (1954–1996 and 1997–2016) in YJG Watershed; The red dashed line is the 0 t ha^{-1} yr^{-1} and negative values indicate a net loss of soil whilst positive values indicate a net gain of soil

of the transect. The sampling variability is high along 160 m transects and because the soil cores have been collected from different points (even if close to the previous ones), it is likely that different estimates of mean annual soil loss have been obtained for each period. A similar situation was mentioned by Loughran and Balog (2006) for cultivated land. However, there is a detectable reduction of the erosion rates in the second time window. The mean soil erosion reduced significantly in the second time window of 1997–2016 (-3.18 ± 0.91 t ha^{-1} yr^{-1}) when compared with the first time window of 1954–1996 (-12.32 ± 3.58 t ha^{-1} yr^{-1}). The reason why soil erosion magnitude is higher in 1954–1996 period than in 1997–2016 period was because the S3 forest slopes were immature in 1996, with low canopy density and poor soil conserving efficiency.

For S4, the spatial variability of soil redistribution patterns along the hillslope are significantly less fluctuating when compared to S3 and also the differences between the two investigated time windows is much smaller. The difference between the mean soil redistribution along the transects during 1954–1996 period (-1.20 ± 1.27 t ha^{-1} yr^{-1}) and 1997–2016 period (-2.11 ± 0.53 t ha^{-1} yr^{-1}) is not statistically significant due to S4 forest slope being sown by aerial seeding in 1965 which has developed into a mature forest with dense canopy and strong soil erosion resistance. The high variation of the soil redistribution at the same sampling points for the two investigated time windows were caused by the high microvariability of ^{137}Cs

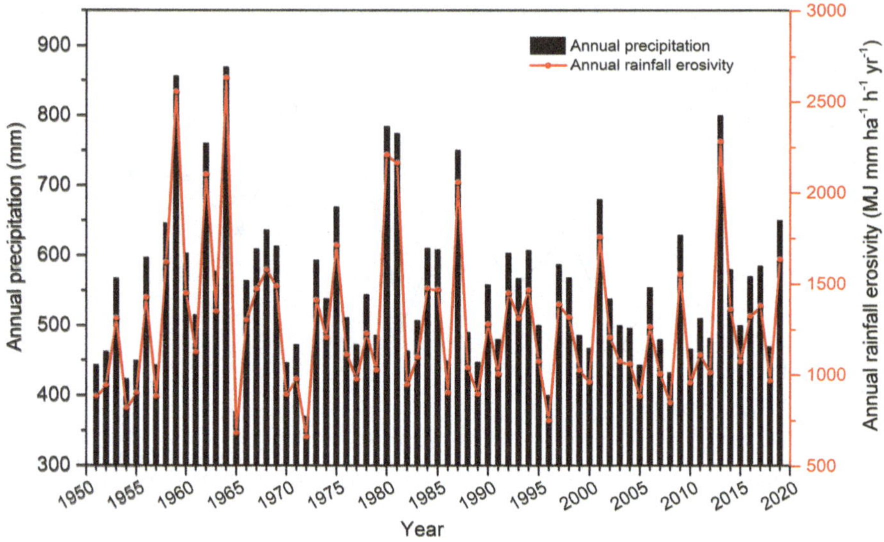

Fig. 2.46 Long term dynamics of annual rainfall and rainfall erosivity since 1951 until 2019 in YJG Watershed

inventory and the different (a distance of up to ca. 2 m) positions of core samples collected during the first and second sampling campaigns.

The annual rainfall amount was collected from Yan'an Meteorological Bureau monitoring station which is 9.7 km far from the YJG Watershed. Average annual rainfall erosivity (R) was estimated using the rainfall erosivity factor (R-factor) used for Universal Soil Loss Equation (USLE), which was calculated by the formula of Renard (Renard and Freimund 1994). Figure 2.46 presents annual values of the rainfall and R-factor between 1951 and 2019 in YJG Watershed.

The average annual rainfall was 559 ± 17 mm during first time window (1951–1996) and 542 ± 23 mm during second time window (1997–2019) and R-factor was 1308 ± 67 $MJ{\cdot}mm{\cdot}ha^{-1}{\cdot}h^{-1}{\cdot}yr^{-1}$ (1951–1996) and 1234 ± 86 $MJ{\cdot}mm{\cdot}ha^{-1}{\cdot}h^{-1}{\cdot}yr^{-1}$ (1997–2019). These differences between the two investigated periods for both rainfall and R-factor are not significant. This is consistent with soil redistribution in S2 sloping farmland where the land use did not experience any change over the last 100 years, i.e., the difference of the mean soil redistribution along the transects between 1954–1996 and 1997–2019 periods were not statistically significant. Both forested slopes (S3 and S4) have lower erosion rates than cultivated S2 slope. On S3 forest hillslope (planted by aerial seeding in 1983) the erosion was significantly higher during the first period (1954–1996) and lower during the second time window (1996–2016). This is because during the first period the forest was still young and provided weaker protection against the erosion than during the second period when the forest was already more dense. On the S4 forest hillslope (planted by aerial seeding in 1965), there was no significant change between the two peiods, because here the forest was planted earlier and it was mature alredy during the first period. These results indicate that the impact of climate change did not manifest by different soil erosion rates between the two studied time windows and the major differentiation factor is land use change. At those sites, where the land use was stable the erosion rates remained similar during both investigated time windows.

2.4 Contribution from Africa: Use of ^{137}Cs Resampling Approach to Assess Changing Soil Erosion and Deposition Rates in Morocco

Most of the studies carried out in the past in Morocco to quantify soil erosion and deposition rates have been employing the conventional techniques such as experimental runoff plots or empirical erosion models. Nevertheless, the potential of using fallout radionuclides (FRNs), especially Cesium-137 (^{137}Cs), excess Lead-210 ($^{210}Pb_{ex}$) and Beryllium-7 (^{7}Be) for soil erosion and sedimentation studies has been demonstrated during the last 30 years by several successful case studies (Bouhlassa et al. 1995, 2000; Benmansour et al. 2002, 2006, 2011, 2013; Nouira et al. 2003; Damnati et al. 2004).

However, the use of resampling FRN tracers to assess change in soil erosion and deposition rates between two time windows and investigate the erosion development trends was never tested in Morocco, before 2016. This approach represents one of the last advances in investigating temporal changes in soil redistribution rates (Porto et al. 2014, 2016).

The purpose of first FRN resampling studies undertaken in Morocco (Moustakim et al. 2019; Benmansour et al. 2019) was to test the ^{137}Cs resampling approach for assessing change in soil erosion and deposition rates, in relation to climate change and soil conservation practices, at some sites where ^{137}Cs technique was already applied in early 2000s (Damnati et al. 2004; Ibrahimi 2005). These study areas are located in two regions in the northern part of Morocco: Tangier-Tetouan and Rabat-Kénitra (Fig. 2.47). Using the ^{137}Cs resampling approach, the average soil erosion rates were compared between two time windows for each region:

- 1954–2000/2002 and 1954–2017/2018 for the cultivated sites in Tangier-Tetouan Region.
- 1954–2007 and 1954–2019 for the forest site in Rabat-Kénitra Region.

In total 6 study sites and 4 reference sites were investigated:

- 5 cultivated sites (HA1, HA2, NA1, RA1, RA2) and 3 reference sites (R1, R2, R3) situated in three watersheds: El Hachef (HA1, HA2, R1), Nakhla (NA1, R2) and Raouz (RA1, RA2, R3) in theTangier-Tetouan Region.
- 1 forest site (MA1 and 1 reference site (R4) in Mâamora Forest in the Rabat-Kénitra Region.

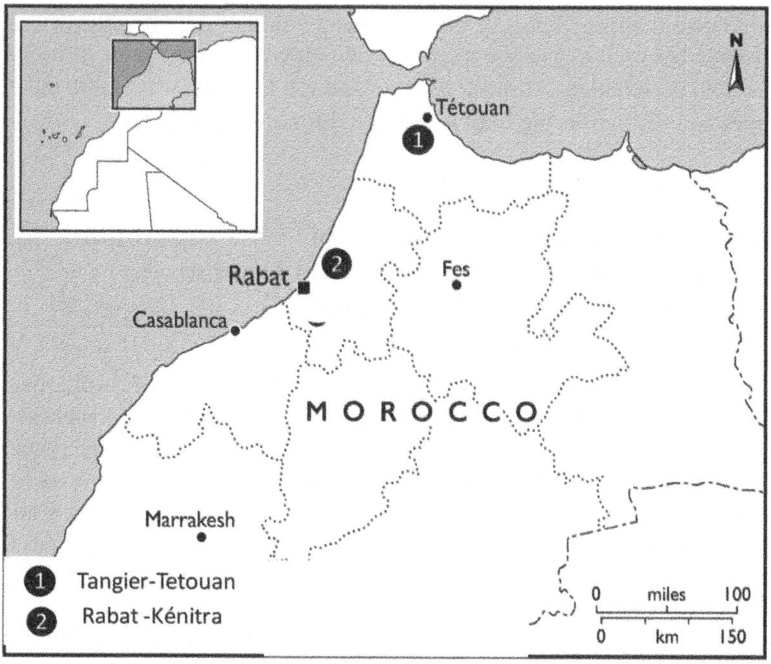

Fig. 2.47 Localization of the regions where the study sites were selected

The Tangier-Tétouan Region is characterized by a subhumid Mediterranean climate with hot dry summer, wet winter and annual precipitation ranging between 600 and 800 mm (Damnati et al. 2004; Ibrahimi 2005; Damnati et al. 2013). The mean annual rainfall reaches 773, 741 and 663 mm for El Hachef, Nakhla and Raouz watersheds respectively. The variation of the annual rainfall since 1978/1979 untill 2015/2016 for the three watersheds show two periods of heavy rains, recording extreme events, that had occurred since 1995 until 1998 and since 2008 until 2011 (Fig. 2.48). The investigated sites were selected in mountainous areas and have steep slopes (11% and 30%). The dominant soil texture for all selected fields is clay and silt (Ibrahimi 2005; Damnati et al. 2013). The soils occupying these fields are characterized by a low calcium carbonate ($CaCO_3$) content that does not exceed 1% (Ibrahimi 2005).

The investigated agricultural fields were generally cultivated in the past, before 2002 cereals (wheat or barley) in rotation with legumes were grown there using conventional tillage. In the last fifteen years (2002–2017), a change in agricultural practices has occurred in all fields (HA1, HA2, NA1, and RA1 except for one (RA2) fallows combined with natural vegetation became to be used more frequently. In addition, this land use was associated with natural olive trees for the field NA1 of Nakhla watershed. Reference sites R1, R2 and R3 were identified near to the study fields of El Hachef, Nakhla and Raouz watersheds, respectively. They are occupied mostly by mosaic of open forests, forest-steppe, shrub and grassland ecosystems with abundant cork oak groves.

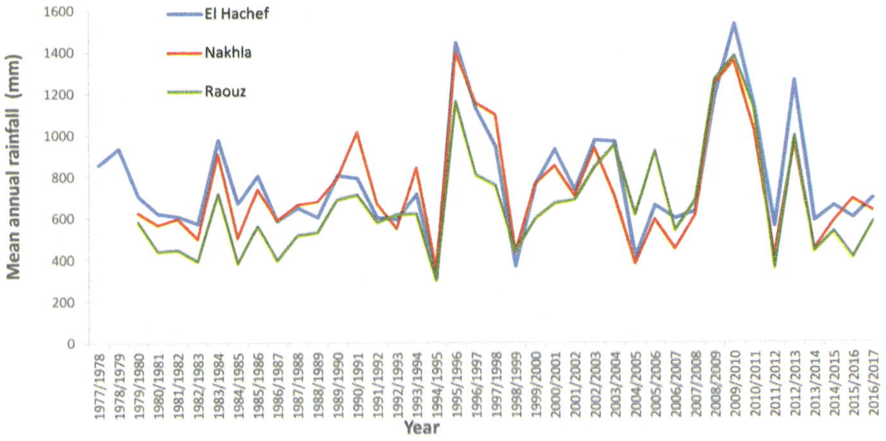

Fig. 2.48 Rainfall dynamics (1977/1978–2015/2016) in El Hachef, Nakhla and Raouz watersheds

The Rabat-Kénitra Region is characterized by a semiarid climate with annual precipitation lower than 500 mm. The studied site belongs to Maamora Forest with much lower tree density as compared to the past. This field is under cork oak grove. It has a slope inclination of about 12% and the mean annual precipitation is 480 mm.

The soil samples were collected to a depth of 40–50 cm using motorised Soil Column Cylinder Auger. The sampling strategy was based on a transect approach (1–4 transects) at studied sites and grid approach at the reference sites. Both bulk and sectioned soil cores were collected for obtaining inventories and ^{137}Cs vertical distributions in the soil. Between 8 and 36 soil samples were collected for each field and 10 samples were collected for each reference site.

All soil samples were firstly dried (~80 °C), lightly ground, sieved (<2 mm) and the fine fraction was well homogenized with a mixer to ensure that the subsample taken for analysis was representative of the total mass of the fine fraction. The ^{137}Cs was measured by gamma spectrometry using two high resolution coaxial hyper-pure germanium detectors (p-type) with relative efficiency between 30 and 34%. The counting rates varied from 12 to 24 h providing a precision of about 6 to 12% at 95% level of confidence. Under these conditions, the detection limits were in the range of 0.20–0.40 Bq kg^{-1} for ^{137}Cs.

The vertical ^{137}Cs distribution associated with the reference and agricultural sites of Tangier-Tétouan and Rabat-Kénitra Regions is in accordance with typical shapes of ^{137}Cs profiles found in literature. Indeed, the shapes of the ^{137}Cs profiles generally show exponential ^{137}Cs decrease with the depth in all reference sites located in El Hached, Raouz and Nakhla watersheds (Fig. 2.49).

This phenomenon produces a nearly constant ^{137}Cs concentration in the upper 3–5 cm with a slight peak (Walling and Quine 1995; Benmansour et al. 2013). On the other hand, according to some authors, the vertical distribution of ^{137}Cs in the soil profile can be divided into three types: exponential, peak and decreasing type (Walling and Quine 1992; Du et al. 1998). Generally, the depth distribution of ^{137}Cs

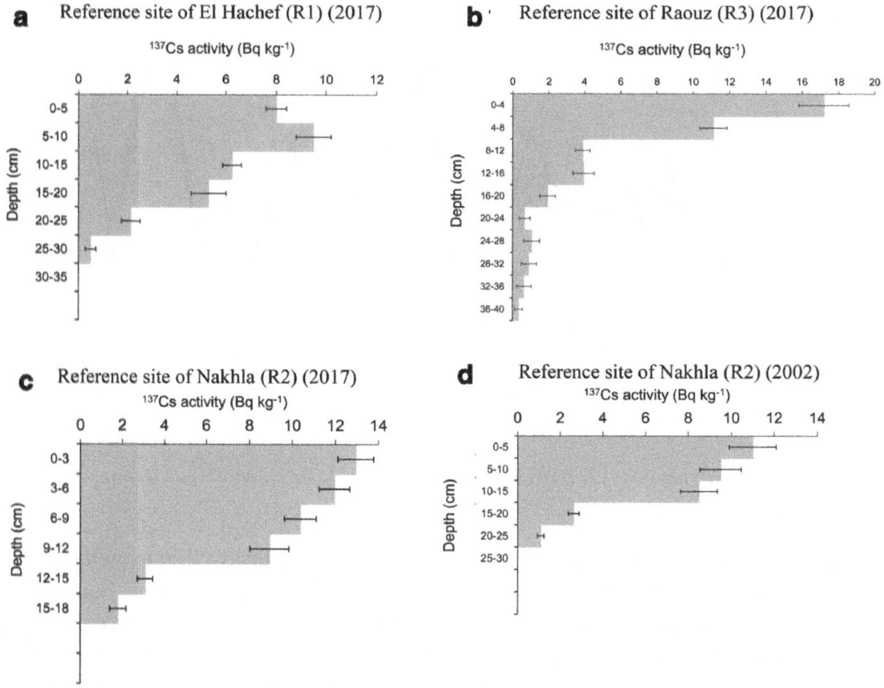

Fig. 2.49 Exponentially decreasing ^{137}Cs vertical distributions at reference sites of El Hachef (**a**), Raouz (**b**) and Nakhla (**c**, **d**)

depends mainly on soil type and soil structure differing in the clay and organic matter distribution (Du et al. 1998).

Therefore, the ^{137}Cs vertical distribution associated with undisturbed sites showed different shapes in the upper soil layers with certain dependence on soil particle size, although the exponential model associated with bioturbation process observed near the soil surface, is the most universal type of ^{137}Cs profile for an uncultivated soil, as pointed out by many authors (Zhang et al. 1990; Porto et al. 2003; Soto and Navas 2004).

Almost the same ^{137}Cs depth profile shapes, like those found recently, were obtained during the previous sampling campaigns as indicated in Fig. 2.49 showing a comparison between ^{137}Cs vertical distributions of the soil profiles sampled in 2002 and 2017 at Nakhla reference site (NA1). Indeed, the ^{137}Cs activity decreased with depth and more than 90% of ^{137}Cs was retained in the upper 0–20 cm of the soil profile.

For the cultivated fields, the ^{137}Cs depth distribution is almost homogeneous over the depth of ploughed horizon due to its periodical mixing (and consequent homogenization) caused by cultivation practices (Fig. 2.50).

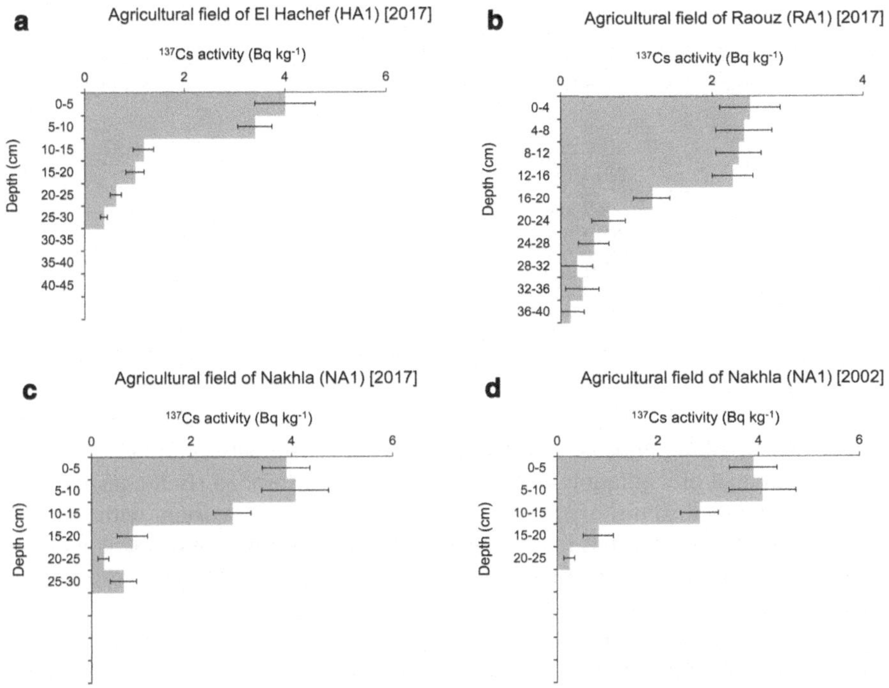

Fig. 2.50 Exponentially decreasing ^{137}Cs vertical distributions at agricutural sites of El Hachef (**a**), Raouz (**b**) and Nakhla (**c**, **d**)

The plough layer thickness is ranging between 10 and 20 cm depending on the agricultural practice adopted in the investigated field. A similar shape of ^{137}Cs depth distribution (Fig. 2.50) in the cultivated site was obtained in 2002, but the ^{137}Cs depth seems to be slightly lower than the one found in 2017. Similar behaviours of ^{137}Cs were observed on agricultural land in other studies carried in Morocco (Benmansour et al. 2013; Damnati et al. 2013).

The ^{137}Cs reference inventories with their standard deviations were reported in Table 2.4 for all sampled reference sites.

Three values of ^{137}Cs inventories were given for each reference site: value of first sampling, value of second sampling and the value of first sampling decay corrected to the time of the second sampling. They are found in the range of 1709–2142 Bq m^{-2} and 1251–1710 Bq m^{-2} for the first and second sampling campaign, respectively. The coefficients of variation are in the range of 12–24% and 15–17% for the first and second sampling, respectively. The obtained results showed that the recent ^{137}Cs inventories are very close to the decay corrected ^{137}Cs inventories of first sampling confirming the stability of the selected reference sites during the first and second periods (1954–2000/2002 and 1954–2017/2018 for Tangier-Tétouan Region and 1954–2007 and 1954–2019 for Rabat-Kénitra Region).

Table 2.4 Reference inventories of ^{137}Cs from first and second sampling

Watershed/area	Reference site	Years of sampling	First ^{137}Cs inventory (Bq m^{-2})	Second ^{137}Cs inventory (Bq m^{-2})	Decay corrected first ^{137}Cs inventory (Bq m^{-2})
El Hachef	R1	2000 and 2017	2542 ± 540	1685 ± 321	1707 ± 357
Nakhla	R2	2002 and 2017	2142 ± 252	1517 ± 223	1520 ± 179
Raouz	R3	2001 and 2018	2651 ± 252	1710 ± 321	1713 ± 157
Maâmora	R4	2007 and 2019	1709 ± 420	1251 ± 187	1328 ± 321

In addition, taking into account the relation established between ^{137}Cs inventory and mean annual precipitation in different regions of Morocco (Benmansour et al. 2013), the obtained values of ^{137}Cs inventories seem to be consistent with previous data (Fig. 2.51). Indeed, the recent values are found in good agreement with the straight line representing the fitting curve of the previous ^{137}Cs data versus mean annual rainfall (Fig. 2.51).

The ^{137}Cs inventories from second sampling in 2017–2018, associated with all individual points of the investigated fields (HA1, HA2, NA1, RA1, RA2), located in the Tangier-Tetouan Region, are between 75 and 2324 Bq m^{-2} while the inventories of these fields recorded in first sampling in 2001–2002 ranged between 196 and 3422 Bq m^{-2}. For the Rabat-Kénitra Region, the recent inventories recorded in 2007 associated with all individual points of the field MA1 located in Maamora Forest are between 508 and 1650 Bq m^{-2} while the previous inventories recorded in 2019 ranged between 825 and 2500 Bq m^{-2}. Table 2.5 summarizes the mean ^{137}Cs inventories

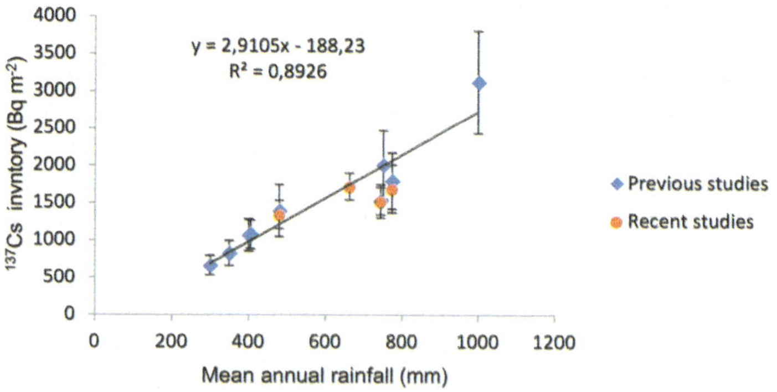

Fig. 2.51 Relationship between mean annual precipitation and ^{137}Cs inventories of the reference sites of previous studies (Bouhlassa et al. 1995, 2000; Benmansour et al. 2002, 2006, 2011, 2013; Nouira et al. 2003; Damnati et al. 2004) (blue rectangular marks) and the recent measurements of the second sampling (orange bullet marks)

accompanied with their respective uncertainties, obtained for each field associated with previous and recent periods.

Table 2.5 shows that all the study sites have mostly been exposed to soil erosion over the investigated periods of 1954–2002 and 1954–2017 in comparison with the reference sites (Table 2.4). On the other hand, the results indicated that the ^{137}Cs inventories from the second sampling of all fields (HA1, HA2, RA1, RA2, MA1), except for the NA1, were significantly lower than the inventories from the first sampling as a result of radioactive decay and soil erosion. For the NA1 Field, the ^{137}Cs inventories for both periods are almost the same, what means that during the second time window (2002–2017), this field was stable and no significant removal of ^{137}Cs took place there (Moustakim et al. 2019). This stabilization could be explained by the positive impact of improved agricultural practices using more frequently fallows combined with natural vegetation and olive plantations in addition to the traditional cultivation based on legume-cereal crop rotation.

The spatial variability of ^{137}Cs inventory within the field has also been examined for fields where at least two transects or more were sampled. This variability, which can exceed 20%, could be explained mainly by the slope variability or micro-variability and vegetation variability. Generally, if there is a decrease of the slope with the distance from the top to the bottom of the field, low ^{137}Cs activities, which are associated with high soil erosion rates, are observed in the upslope areas, while high activities, associated with lower erosion or deposition rates are observed in the downslope areas (Benmansour et al. 2013).

As an example, Fig. 2.52 shows the ^{137}Cs inventory values for different points along four transects (T1, T2, T3, T4) from the upslope to the downslope of the NA1 Field obtained in 2002 (Fig. 2.52a) and 2017 (Fig. 2.52b).

Both sets of ^{137}Cs inventories derived from the twenty samples provided by the first and the resampling campaigns along the four parallel transects are substantially lower than the reference inventory. However, it is noticeable that the highest ^{137}Cs concentrations that were observed in the downslope of four transects (Fig. 2.52a) sampled in 2002 were displaced to their middle part (Fig. 2.52b) in 2017. This suggests that the middle part of the field was less eroded than the downslope part in 2017 (Moustakim et al. 2019). This result could be attributed to the change in land use between both periods. This resulted to a new redistribution of ^{137}Cs inventories

Table 2.5 Previous and recent mean ^{137}Cs inventories at the study sites

Watershed/area	Field	First period	^{137}Cs inventory (Bq m^{-2})	Second period	^{137}Cs inventory (Bq m^{-2})
El Hachef	HA1	1954–2001	1278 ± 128	1954–2017	886 ± 76
	HA2	1954–2001	1470 ± 147	1954–2017	440 ± 47
Nakhla	NA1	1954–2002	652 ± 65	1954–2018	666 ± 71
Raouz	RA1	1954–2001	1558 ± 156	1954–2018	1014 ± 95
	RA2	1954–2001	2283 ± 228	1954–2018	1541 ± 154
Mâamora	MA1	1954–2007	1519 ± 117	1954–2019	1106 ± 75

Fig. 2.52 Distribution of ^{137}Cs activity along 4 transects (T1, T2, T3, T4) on Nakhla (NA1) and ^{137}Cs reference line (REF) recorded in 2002 (a) and 2017 (b)

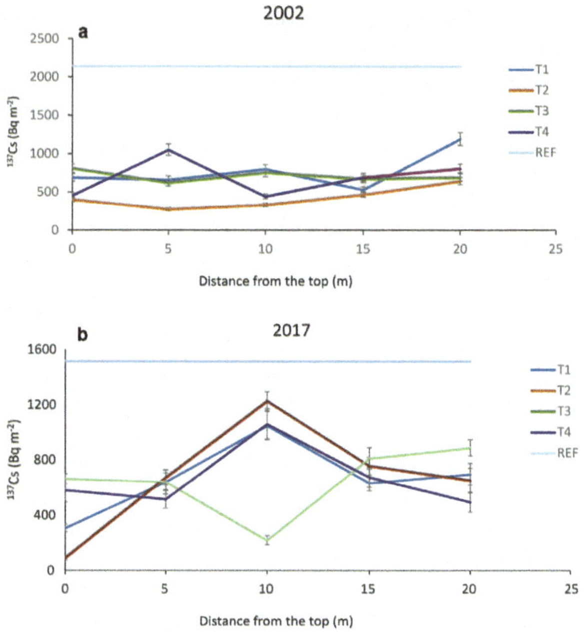

within the field. Indeed, since 2002, the NA1 Field was subject to changes in land management introducing more frequent fallows combined with natural vegetation and olives.

Mass Balance Model II (MB2) and Profile Distribution Model (PDM) were used to calculate the soil erosion and deposition rates at cultivated fields located in Tangier-Tétouan Region and the non-cultivated field (degraded forest) in Rabat-Kénitra Region. To avoid any shift of soil erosion between first and second periods due to change in ^{137}Cs reference inventory, the values obtained by the first samplings were also used for the second samplings after radioactive decay correction.

Table 2.6 summarizes the results of soil redistribution rates corresponding to all investigated fields (HA1, HA2, NA1, RA1, RA2, MA1) in the first time window and the second time window.

Therefore, the recent gross soil erosion rates, over the 1954–2017/2018 period, associated with the agricultural fields of Tangier-Tétouan Region, ranged between 10.3 and 42.7 t ha^{-1} yr^{-1} while the gross deposition rates were found low, between 0.5 and 4.6 t ha^{-1} yr^{-1}. The difference between the obtained values reflects the difference of many parameters including land use, slope and soil proprieties (soil texture and organic matter). Based on the values of gross soil erosion and deposition rates, the net soil erosion rates were calculated to 5.7–42.7 t ha^{-1} yr^{-1} indicating that the sediment delivery ratios for most of the fields, except for the RA2 Field, are almost 100%. All eroded soil material is exported beyond the fields due to generally steep slopes generating great transporting capacity of surface runoff and the lack of

Table 2.6 Soil redistribution rates (t ha^{-1} yr^{-1}) and sediment delivery ratio of studied fields (FP—First period; SP—Second period)

Field	Gross erosion Rate		Gross deposition rate		Net erosion rate		Sediment delivery ratio	
	FP	SP	FP	SP	FP	SP	FP (%)	SP (%)
HA1	26.3	20.7	0.0	0.5	26.3	20.2	100	97
HA2	50.8	42.7	0.0	0.0	50.8	42.7	100	100
NA1	36.1	29.0	0.0	0.0	36.1	29.0	100	100
RA1	19.4	16.1	0.0	0.0	19.4	16.1	100	100
RA2	8.5	10.3	4.0	4.6	4.5	5.7	53	55
MA1	7.3	8.7	2.5	1.4	4.8	6.9	66	80

natural vegetation which would protect the soil. For the RA2 Field, the sediment delivery ratio was about 55%.

The values of first time window (1954–2000/2002) for the HA1, HA2, NA1, RA1 and RA2 Fields were between 8.5 and 50.8 t ha^{-1} yr^{-1} for the gross erosion rates and between 0.0 and 4.0 t ha^{-1} yr^{-1} for the gross deposition rates. The net soil erosion rates were found the same as the gross erosion rates except for the last RA2 Field (Table 2.6).

For the degraded field overgrown by forest (MA1) at Maâmora Site in Rabat-Kénitra Region the gross soil erosion rate over the whole study period (1954–2019) was about 8.7 t ha^{-1} yr^{-1} while the gross soil erosion rate over first time window (1954–2008), was 7.3 t ha^{-1} yr^{-1}. Taking into account the gross soil deposition rates of 1.4 and 2.5 t ha^{-1} yr^{-1} obtained for the second and first period respectively, the net soil erosion rates were 6.7 and 5.2 t ha^{-1} yr^{-1} and the sediment delivery ratios were 80% and 66% for both time windows respectively.

The comparison between both periods for the fields which were subjects to decrease or increase of soil erosion rates is given on Fig. 2.53

The gross soil erosion rates on the HA1, HA2, NA1, RA1 Fields obtained in 2017/2018 are lower than those obtained in 2000/2002 (Fig. 2.53a). They were reduced by 23, 16, 20 and 17%, respectively. This decrease in soil erosion rate from the first to the second period can be explained by the beneficial impact of agricultural practices based on more frequent use of fallows combined with natural vegetation and also with olive plantations in the case of NA1. These techniques were effective in reducing soil erosion on the investigated fields.

Based on the same resampling approach for the 1954–1974 and 1954–1998 periods, a study conducted on a cultivated field in southern Iowa, USA found that reduced soil loss and greater deposition were attributed to the introduction of grass buffer strips and conservation tillage (Fornes et al. 2005). Another study (Loughran and Balog 2006), carried out in a vineyard in Australia using ^{137}Cs resampling (samples collected in 1984–1985 and 2004) demonstrated a reduction in soil erosion as a result of the incorporation of subsoil into the plough layer, and the cessation of clean till cultivation. In the study reported by Porto et al. (2014), both estimates

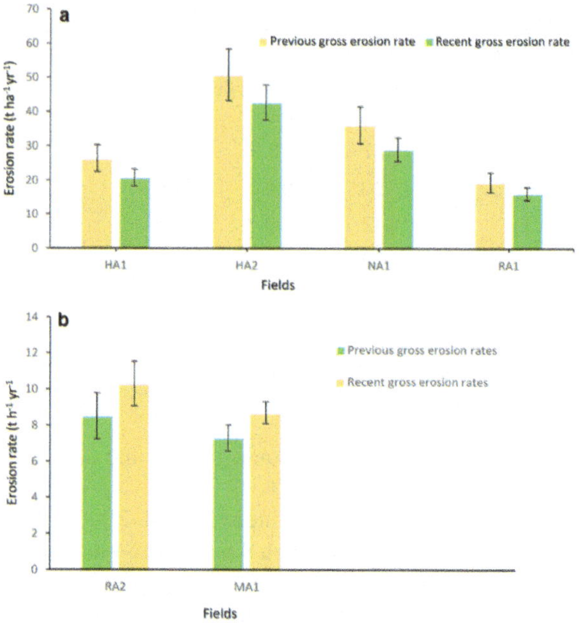

Fig. 2.53 Decrease (**a**) or increase (**b**) of soil erosion rates within the study fields between the first and the second period

of mean annual soil loss for the periods 1954–1998 and 1999–2013 using ^{137}Cs resampling technique and the measurements of mean annual sediment yield, provide similar evidence of a lack of change in soil redistribution.

On RA2 and MA1, which were not affected by land use change between the two studied periods, the soil erosion rate was higher during the second time window. The gross erosion rates increased by 21% and 19% and the net erosion rates by 28% and 45%, respectively. This increase could be attributed to climate change impact on soil erosion, although further investigations and more data are needed to confirm these results. During the second time window the extreme rainfall events were recorded (Fig. 2.48). These rainfalls could be responsible for the increase of soil erosion rates. Indeed, the climate change endangers the sustainability of agroecosystems in the Rif Mountains (Sadiki et al. 2007). In Russia, according to long-term monitoring data from agricultural watersheds, the extreme precipitation contributed by 70–80% of total soil loss (Sadiki et al. 2007) whereas, according to Edwards and Owens (1991), an increase of 20% in the average erosion rate was simulated in response to a 7% decrease in annual precipitation. However, sustainable land use and beneficial agricultural practices could minimize climate change impact on soil stability. In this regard, other studies have shown that the impacts of land use change on soil erosion may be more prominent than the climate change impact (Edwards and Owens 1991; Mullan et al. 2012; Paroissien et al. 2015).

In conclusion, this work highlighted the potential of using the ^{137}Cs resampling approach in Morocco to assess changes in soil erosion rates in relation to climate change and the effectiveness of soil conservation practices in the Tangier-Tétouan and Rabat-Kénitra Regions. The main findings can be summarized as follows:

The ^{137}Cs resampling approach showed a decrease in soil erosion rates by 16–23% on four agricultural studied fields between two periods, when traditional soil conservation practices were adopted on these fields during the second period. These practices were based on more frequent use of fallow combined with natural vegetation and olive plantations;

For two fields, one agricultural and one with degraded forest, with no change in land use, the ^{137}Cs resampling technique showed increases of soil erosion rates by 21% and 19%, which could be explained by the climate change impact on soil loss, due to extreme rainfall events occurred in the second period.

This approach combined with the use of other nuclear and isotopic techniques will be pursued in other Moroccan regions for better exploring of its potential and further understanding of the impacts of climate change and land management on soil erosion to provide useful and reliable information for decision makers.

References

Afshar FA, Ayoubi S, Jalalian A (2010) Soil redistribution rate and its relationship with soil organic carbon and total nitrogen using ^{137}Cs technique in a cultivated complex hillslope in western Iran. J Environ Radioact 101:606–614

Anh PTQ, Gomi T, MacDonald LH, Mizugaki S, Van Khoa P, Furuichi T (2014) Linkages among land use, macronutrient levels, and soil erosion in northern Vietnam: a plot-scale study. Geoderma 232–234:352–362

Bazzoffi P, Panicucci M (1983) Erosione sui versanti e conseguente sedimentazione in piccoli serbatoi artificiali. Annali Ist Sper Studio e Difesa Suolo, Firenze 14:127–178

Benmansour M, Ibn Majah M, Marah H, Marfak T, Walling DE (2002) Use of the ^{137}Cs technique in soil erosion in Morocco—case study of the Zitouna basin in the north. In: International Symposium on Nuclear Techniques in Integrated Plant Nutrients, Water and Soil Management. IAEA Publication, IAEA—CSP11/C, pp 308–315

Benmansour M, Laissaoui A, Benbrahim S, Ibnmajah M, Chafik A, Povinec P (2006) Distribution of anthropogenic radionuclides in Moroccan coastal waters and sediments. Radioact Environ e-Book Ser 8:145–150

Benmansour M, Nouira A, Bouksirat H, Duchemin M, El Oumri M, Mossadek R, Benkdad A, Ibn Majah M (2011) Estimates of long and short-term rates of soil erosion using ^{137}Cs, $^{210}Pb_{ex}$ and ^{7}Be measurements: case study of one agricultural field in semi-arid West Morocco. In: Impact of Soil Conservation Measures on Erosion Control and Soil Quality. IAEA-TECDOC-1665, pp 159–174

Benmansour M, Mabit L, Nouira A, Moussadek R, Bouksirate H, Duchemin M, Benkdad A (2013) Assessment of soil erosion and deposition rates in a Moroccan agricultural field using fallout ^{137}Cs and $^{210}Pb_{ex}$. J Environ Radioact 115:97–106. https://doi.org/10.1016/j.jenvrad.2012.07.013

Benmansour M, Moussadek M, Moustakim M, Zouagui A, Nouira A, Damnati B, Iaaich H, Mrabet R, Fulajtar E, Mabit L (2019) Use of fallout radionuclides to assess change in soil erosion and sedimentation rates in Northern Morocco. Proceeding of Global Symposium on Soil Erosion, 15-17th May, 2019, Food and Agriculture Organisation of the United Nations, Rome, 656–660.

Bernard C, Laverdiere MR (1992) Spatial redistribution of Cs-137 and soil erosion on Orleans Island. Quebec Can J Soil Sci 72:543–554

Bouhlassa S, Azenfar A, Machrouh A (1995) ^{137}Cs fallout as tracer of erosion and sedimentation in big catchment. Appl Radiat Isot 46(6–7):659–660

Bouhlassa S, Moukchane M, Aiachi A (2000) Estimates of soil erosion and deposition of cultivated soil of Nakhla watershed, Morocco, using ^{137}Cs technique and calibration models. Acta Geol Hisp 35(3–4):239–249

Campbell BL, Loughran RJ, Elliott GL (1982) Caesium-137 as an indicator of geomorphic processes in a drainage basin system. Aust Geogr Stud 20:49–64

Campbell BL, Loughran RJ, Elliott GL, Shelly DJ (1986) Mapping drainage basin sediment sources using caesium-137. Int Assoc Hydrol Sci 159:437–446

Cao YZ, Coote DR, Rees HW, Wang C, Chow TL (1994) Effects of intensive potato production on soil quality and yield at a benchmark site in New Brunswick. Soil Tillage Res 29:23–34

Chappell A, Sanderman J, Thomas M, Read A, Leslie C (2012) The dynamics of soil redistribution and the implications for soil organic carbon accounting in agricultural south-eastern Australia. Glob Change Biol 18(6):2081–2088

Chappell A, Li Y, Yu HQ, Zhang YZ, Li XY (2015) Cost-effective sampling of ^{137}Cs-derived net soil redistribution: part 2—estimating the spatial mean change over time. J Environ Radioact 144:168–174

Damnati B, Ibrahimi S, Radakovitch O (2004) Utilisation du ^{137}Cs pour l'estimation des taux d'érosion dans un bassin versant du nord du Maroc. Sécheresse 15(2):195–199

Damnati B, Ibrahimi S, Benhardouze O, Radakovitch O, Moukhchane M (2010) Quantification de l'érosion par le ^{137}Cs et le ^{210}Pb en excès: cas d'une parcelle au niveau du bassin versant du barrage Nakhla (Région de Tétouan, Nord-ouest du Maroc). Afr Geosci Rev 17(3):215–223

Damnati B, Ibrahimi S, Radakovitch O (2013) Quantifying erosion using ^{137}Cs and ^{210}Pb in cultivated soils in three Mediterranean watersheds: synthesis study from El Hachef, Raouz and Nakhla (north west Morocco). J Afr Earth Sci 79:50–57

de Roo AJJ (1991) The use of ^{137}Cs as a tracer in an erosion study in south Limburg (The Netherlands) and the influence of Chernobyl fallout. Hydrol Process 5:215–227

de Jong E, Kachanoski RG (1988) The importance of erosion in the carbon balance of prairie soils. Can J Soil Sci 68:111–119

de Jong E, Villar H, Bettany JR (1982) Preliminary investigations on the use of ^{137}Cs to estimate erosion in Saskatchewan. Can J Soil Sci 62:673–683

de Jong E, Begg CBM, Kachanoski RG (1983) Estimates of soil erosion and deposition for some Saskachewan soils. Can J Soil Sci 63:607–617

de Jong E, Wang C, Rees HW (1986) Soil redistribution on three cultivated New Brunswick hillslopes calculated from ^{137}Cs measurements, solum data and the USLE. Can J Soil Sci 66:721–730

Di Stefano C, Ferro V, Porto P, Rizzo S (2005) Testing a spatially distributed sediment delivery model (SEDD) in a forested basin by caesium-137 technique. J Soil Water Conserv 60(3):148–157

Du M, Yang H, Chang Q, Minami K, Hatta T (1998) Caesium-137 fallout depth distribution in different soil profiles and significance for estimating soil erosion rate. Sci Soils. https://doi.org/10.1007/s10112-998-0003-1

Edwards WM, Owens LB (1991) Large storm effects on total soil erosion. J Soil Water Conserv 1:75–78

Elliott GL, Campbell BL, Loughran RJ (1990) The correlation of erosion measurement and soil caesium-137 content. Appl Radiat Isot 41:713–717

Fornes WL, Whiting PJ, Wilson CG, Matisoff G (2005) Caesium-137—derived erosion rates in an agricultural setting: the effects of model assumptions and management practices. Earth Surf Process Land 30:1181–1189

Fulajtar E (2000) Assessment of soil erosion through the use of ^{137}Cs at Laslovke Bohunice, Western Slovakia. Acta Geol Hisp 35:291–300

Gaspar L, Navas A (2013) Vertical and lateral distributions of ^{137}Cs in cultivated and uncultivated soils on Mediterranean hillslopes. Geoderma 207–208:131–143. https://doi.org/10.1016/j.geoderma.2013.04.034

Gaspar L, Navas A, Walling DE, Machín J, Gómez-Arozamena J (2013a) Using ^{137}Cs and $^{210}Pb_{ex}$ to assess soil redistribution on slopes at different temporal scales. CATENA 102:46–54. https://doi.org/10.1016/j.catena.2011.01.004

Gaspar L, Navas A, Machín J, Walling DE (2013b) Using $^{210}Pb_{ex}$ measurements to quantify soil redistribution along two complex toposequences in Mediterranean agroecosystems, northern Spain. Soil Tillage Res 130:81–90. https://doi.org/10.1016/j.still.2013.02.011

Gaspar L, Quijano L, Lizaga I, Navas A (2019) Effects of land use on soil organic and inorganic C and N at ^{137}Cs traced erosional and depositional sites in mountain agroecosystems. CATENA 181:104058. https://doi.org/10.1016/j.catena.2019.05.004

Gaspar L, Mabit L, Lizaga I, Navas A (2020) Lateral mobilization of soil carbon induced by runoff along karstic slopes. J Environ Manage 260:110091. https://doi.org/10.1016/j.jenvman.2020.110091

Gaspar L, Lizaga I, Navas A (2021) Spatial distribution of fallout and lithogenic radionuclides controlled by soil carbon and water erosion in an agroforestry South-Pyrenean catchment. Geoderma 391:114941. https://doi.org/10.1016/j.geoderma.2021.114941

He Q, Walling DE (1997) The distribution of fallout ^{137}Cs and ^{210}Pb in undisturbed and cultivated soils. Appl Radiat Isot 48:677–690

Higgitt DL, Froehlich W, Walling DE (1992) Application and limitations of Chernobyl radiocaesium measurements in a Carpathian erosion investigation, Poland. Land Degrad Rehabil 3:15–26

Ibrahimi S (2005) Quantification de l'érosion hydrique des sols par deux radioéléments ^{137}Cs et $^{210}Pb_{ex}$ au niveau des bassins versants El Hachef et Raouz (Région de Tanger-Tétouan, Nord du Maroc). Université Abdelmalek Essaadi, Faculté des Sciences et Techniques, Tanger, Maroc, Thèse de doctorat, 259p

Ionita I, Margineanu RM (2000) Application of for measuring soil erosion/deposition rates in Romania. Acta Geol Hisp 35:311–319

Jiang Y, Gao Y, Dong Z, Liu B, Zhao L (2018) Simulations of wind erosion along the Qinghai-Tibet railway in north-central Tibet. Aeolian Res 32:192–201

Kachanoski RG (1987) Comparison of measured soil 137-Cesium losses and erosion rates. Can J Soil Sci 67:199–203

Kachanoski RG (1993) Estimating soil loss from changes in soil cesium-137. Can J Soil Sci 73:629–632

Kachanoski RG, de Jong E (1984) Predicting the temporal relationship between cesium-137 and erosion rate. J Environ Qual 13:301–304

Kachanoski RG, von Bertoldi AP (1996) Monitoring soil loss and redistribution using ^{137}Cs. Green Plan Report. Agriculture and Agri-food Canada, Guelph, 29pp

Kachanoski RG, Protz RD, Gregorich EG (1992) Management of farm field variability: I. Quantification of soil loss in complex topography. SWEEP/TED report. Centre for Soil and Water Conservation. University of Guelph, Guelph, ON

Karbassi AR, Amirnezhad R (2004) Geochemistry of heavy metals and sedimentation rate in a bay adjacent to the Caspian Sea. Int J Environ Sci Technol 1:191–198

Kato H, Onda Y, Tanaka H (2010) Using ^{137}Cs and $^{210}Pb_{ex}$ measurements to estimate soil redistribution rates on semi-arid grassland in Mongolia. Geomorphology 114:508–519. https://doi.org/10.1016/j.geomorph.2009.08.009

Kirchner G, Ehlers H (1998) Sediment geochronology in changing coastal environments: potentials and limitations of the ^{137}Cs and ^{210}Pb methods. J Coastal Res 14(2):483–492

Konz N, Schaub M, Prasuhn V, Bänninger D, Alewell C (2009) Cesium-137-based erosion rate determination of a steep mountainous region. J Plant Nutr Soil Sci 172:615–622

Li Y, Lindstrom MJ (2001) Evaluating soil quality-soil redistribution relationship on terraces and steep hillslope. Soil Sci Soc Am J 65(5):1500–1508

Li Y, Zhang QW, Reicosky DC, Bai LY, Lindstrom MJ, Li L (2006) Using Cs-137 and Pb-210(ex) for quantifying soil organic carbon redistribution affected by intensive tillage on steep slopes. Soil Tillage Res 86(2):176–184

Li S, Lobb DA, Kachanoski RG, McConkey BG (2011) Comparing the use of the traditional and repeated-sampling-approach of the Cs-137 technique in soil erosion estimation. Geoderma 160(3–4):324–335. https://doi.org/10.1016/j.geoderma.2010.09.029

Lizaga I, Quijano L, Gaspar L, Navas A (2018) Estimating soil redistribution patterns with ^{137}Cs measurements in a Mediterranean mountain catchment affected by land abandonment. Land Degrad Dev 29:105–117. https://doi.org/10.1002/ldr.2843

Lobb DA, Kachanoski RG (1999) Modelling tillage erosion in the topographically complex landscapes of southwestern Ontario, Canada. Soil Tillage Res 51:261–277

Lobb DA, Kachanoski RG, Miller MH (1999) Tillage translocation and tillage erosion in the complex upland landscapes of southwestern Ontario, Canada. Soil Tillage Res 51:189–209

Longmore ME, O'Leary BM, Rose CW, Chandica AL (1983) Mapping soil erosion and accumulation with the fallout isotope caesium-137. Austr J Soil Res 21:373–385

Loughran RJ, Balog RM (2006) Re-sampling for soil-caesium-137 to assess soil losses after a 19-year interval in a Hunter Valley vineyard, New South Wales. Australia Geogr Res 44(1):77–86

Loughran RJ, Elliott GL, McFarlane DJ, Campbell BL (2004) A survey of soil erosion in Australia using caesium-137. Austr Geogr Stud 42:221–233

Mabit L, Klik A, Benmansour M, Toloza A, Geisler A, Gertsmann UC (2009) Assessment of erosion and deposition rates within an Austrian agricultural watershed by combining ^{137}Cs, $^{210}Pb_{ex}$ and conventional measurements. Geoderma 150:231–239

Mabit L, Meusburger K, Fulajtar E, Alewell C (2013) The usefulness of ^{137}Cs as a tracer for soil erosion assessment: a critical reply to Parsons and Foster (2011). Earth Sci Rev 127:300–307. https://doi.org/10.1016/j.earscirev.2013.05.008

Mandal D, Giri N, Srivastava P, Sah C, Bhusan R, Naregundi K, Mohan MP, Shrivastava M (2019) ^{137}Cs—a potential environmental marker for assessing erosion-induced soil organic carbon loss in India. Curr Sci 117(5):865–871

McHenry JR, Bubenzer GD (1985) Field erosion estimated from ^{137}Cs activity measurements. Trans Am Soc Agric Eng 28:480–483

McHenry JR, Ritchie JC (1977) Estimating field erosion losses from fallout cesium-137 measurements. IAHS Publ. No. 122:26–33

McHenry JR, Ritchie JC, Gill AC (1973) Accumulation of fallout cesium-137 in soils and sediments in selected watersheds. Wat Resour Res 9:676–686

Meusburger K, Mabit L, Park JH, Sandor T, Alewell C (2013) Combined use of stable isotopes and fallout radionuclides as soil erosion indicators in a forested mountain site, South Korea. Biogeosci 10:5627–5638

Mitchell JK, Bubenzer GD, McHenry JR, Ritchie JC (1980) Soil loss estimates from fallout caesium-137 measurements. In: DeBoodt M, Gabriels D (eds) Assessment of erosion. Wiley and Sons, Chichester, pp 393–401

Moustakim M, Benmansour M, Zouagui A, Nouira A, Benkdad A, Damnati B (2019) Use of caesium-137 re-sampling and excess lead-210 techniques to assess changes in soil redistribution rates within an agricultural field in Nakhla watershed. J Afr Earth Sci 156:158–167. https://doi.org/10.1016/j.jafrearsci.2019.04.017

Mullan DJ, Favis-Mortlock DT, Fealy R (2012) Addressing key limitations associated with modelling soil erosion under the impacts of future climate change. Agric Forest Meteorol 156:18–30

Navas A, Walling D (1992) Using caesium-137 to assess sediment movement in a semiarid upland environment in Spain. In Erosion, debris flows and environment in mountain regions. In Walling DE, Davies TR, Hasholt B (eds) International Association Hydrological Sciences (IAHS) Publ. n° 209, 129–138. ISBN 0-947571-38-8. https://iahs.info/uploads/dms/9000.129-138-209-Navas.pdf

Navas A, Machín J, Soto J (2005) Assessing soil erosion in a Pyrenean mountain catchment using GIS and fallout ^{137}Cs. Agric Ecosyst Environ 105:493–506

Navas A, Walling DE, Quine T, Machín J, Soto J, Domenech S, López-Vicente M (2007) Variability in ^{137}Cs inventories and potential climatic and lithological controls in central Ebro valley. Spain. J Radioanal Nucl Chem 274(2):331–339. https://doi.org/10.1007/s10967-007-1119-8

Navas A, López-Vicente M, Gaspar L, Machín J (2013) Assessing soil redistribution in a complex karst catchment using fallout ^{137}Cs and GIS. Geomorphology 196:231–241. https://doi.org/10.1016/j.geomorph.2012.03.018

Navas A, López-Vicente M, Gaspar L, Palazón L, Quijano L (2014) Establishing a tracer based sediment budget to preserve wetlands in Mediterranean mountain agroecosystems (NE Spain). Sci Total Environ 496:132–143. https://doi.org/10.1016/j.scitotenv.2014.07.026

Navas A, Quine TA, Walling DE, Gaspar L, Quijano L, Lizaga I (2017) Relating intensity of soil redistribution to land use changes in abandoned Pyrenean fields using fallout caesium-137. Land Degrad Dev 28:2017–2029. https://doi.org/10.1002/ldr.2724

Nouira A, Sayouty EH, Benmansour M (2003) Use of ^{137}Cs technique for soil erosion study in the agricultural region of Casablanca in Morocco. J Environ Radioact 68(1):11–26

Olivares RU, Bulos ADM, Sombrito EZ (2016) Environmental assessment of soil erosion in Inabanga watershed (Bohol, Philippines). Energy, Ecol Environ 1:98–108

Owens PN (1994) Towards improved interpretation of Caesium-137 measurements in soil erosion studies. Ph.D. Thesis, University of Exeter, UK

Paroissien JB, Darboux F, Couturier A, Devillers B, Mouillot F, Raclot D, Le Bissonnais Y (2015) A method for modeling the effects of climate and land use changes on erosion and sustainability of soil in a Mediterranean watershed (Languedoc, France). J Environ Manag 150:57–68

Parsons AJ, Foster IDL (2011) What can we learn about soil erosion from the use of ^{137}Cs? Earth Sci Rev 108:101–113

Ping Y, Dong Z, Dong G, Zhang X, Zhang Y (2001) Preliminary results of using ^{137}Cs to study wind erosion in the Qinghai-Tibet Plateau. J Arid Environ 47(4):443–452

Porto P, Callegari G (2022) Comparing long-term observations of sediment yield with estimates of soil erosion rate based on recent ^{137}Cs measurements. Results from an experimental catchment in Southern Italy. Hydrol Processes 36. DOI:0.1002/hyp.14663

Porto P, Walling DE (2012a) Validating the use of ^{137}Cs and $^{210}Pb_{ex}$ measurements to estimate rates of soil loss from cultivated land in southern Italy. J Environ Radioact 106:47–57

Porto P, Walling DE (2012b) Using plot experiments to test the validity of mass balance models employed to estimate soil redistribution rates from ^{137}Cs and $^{210}Pb_{ex}$ measurements. Appl Radiat Isot 70:2451–2459

Porto P, Walling DE, Callegari G, Catona F (2006). Using fallout lead-210 measurements to estimate soil erosion in three small catchments in Southern Italy. Water, Air, and Soil Pollution: Focus 6: 657–667

Porto P., Walling D.E., Callegari G. (2009) Investigating the effects of afforestation on soil erosion and sediment mobilisation in two small catchments in Southern Italy. Catena 79:181–188.

Porto P, Walling DE, Ferro V (2001) Validating the use of caesium-137 measurements to estimate soil erosion rates in a small drainage basin in Calabria, Southern Italy. J Hydrol 45:817–832

Porto P, Walling DE, Ferro V, Di Stefano C (2003) Validating erosion rate estimates by caesium-137 measurements for two small forested catchments in Calabria, Southern Italy. Land Degrad Dev 14:389–408

Porto P, Walling DE, Callegari G (2004) Validating the use of caesium-137 measurements to estimate erosion rates in three small catchments in Southern Italy. IAHS Publ 288:75–83

Porto P, Walling DE, Callegari G (2011) Using ^{137}Cs measurements to establish catchment sediment budgets and explore scale effects. Hydrol Processes 25:886–900

Porto P, Walling DE, Alewell C, Callegari G, Mabit L, Mallimo N, Meusburger K, Zehringer M (2014) Use of a ^{137}Cs re-sampling technique to investigate temporal changes in soil erosion and sediment mobilisation for a small forested watershed in southern Italy. J Environ Radioact 138:137–148

Porto P, Walling DE, Cogliandro V, Callegari G (2016a) Exploring the potential for using $^{210}Pb_{ex}$ measurements within a re-sampling approach to document recent changes in soil redistribution rates within a small watershed in southern Italy. J Environ Radioact 164:158–168

Porto P, Walling DE, La Spada C, Callegari G (2016b) Validating the use of ^{137}Cs measurements to derive the slope component of the sediment budget of a small catchment in southern Italy. Land Degrad Dev 27:798–810

Porto P, Walling DE, Callegari G (2018) Using repeated ^{137}Cs and $^{210}Pb_{ex}$ measurements to establish sediment budgets for different time windows and explore the effect of connectivity on soil erosion rates in a small experimental catchment in Southern Italy. Land Degrad Dev 29:1819–1832

Quine T, Navas A, Walling DE, Machín J (1994) Soil erosion and redistribution on cultivated and uncultivated land near Las Bardenas in the Central Ebro River Basin, Spain. Land Degrad Rehabil 5:41–55. https://doi.org/10.1002/ldr.3400050106

Quine TA, Govers G, Walling DE, Zhang X, Desmet PJJ, Zhang Y, Vandaele K (1997) Erosion processes and landform evolution on agricultural land—new perspectives from caesium-137 measurements and topographic-based erosion modelling. Earth Surf Proc and Land 22:799–816

Renard KG, Freimund JR (1994) Using monthly precipitation data to estimate the R-factor in the revised USLE. J Hydrol 157:287–306

Ritchie JC, McCarty GW (2008) Redistribution of soil and soil organic carbon on agricultural landscapes. Sediment dynamics in changing environments. Proceedings of a Symposium Held in Christchurch, New Zealand. IAHS Publications, p 325

Ritchie JC, McHenry JR, Gill AC, Hawks PH (1970) The use of fallout cesium-137 as a tracer of sediment movement and deposition. Proc Miss Water Resour Conf 1970:149–162

Ritchie JC, Spraberry JA, McHenry JR (1974) Estimating soil erosion from the redistribution of fallout ^{137}Cs. Soil Sci Soc Amer Proc 38:137–139

Rodriguez-Lloveras X, Buytaert W, Benito G (2016) Land use can offset climate change induced increases in erosion in Mediterranean watersheds. CATENA 143:244–255

Ruecker GR, Park SJ, Brunner AC, Vlek PLG (2008) Assessment of soil redistribution on two contrasting hillslopes in Uganda using Caesium-137 modelling. Erdkunde 62:259–272

Sadiki A, Faleh A, Navas A, Bouhlassa S (2007) Assessing soil erosion and control factors by the radiometric technique in the Boussouab watershed, Eastern Rif. Morocco. Catena 71(1):13–20

Sankar M, Mishra PK, Green SM (2018) Nationwide soil erosion assessment in India using radioisotope tracers Cs-137 and Pb-210: the need for fallout mapping. Current Sci 115(3):388–390

Schuller P, Walling DE, Sepulveda A, Trumper RE, Rouanet JL, Pino I, Castillo A (2004) Use of ^{137}Cs to estimate changes in soil erosion rates associated with changes in soil management practices on cultivated land. Appl Radiat Isot 60:759–766. https://doi.org/10.1016/j.apradiso.2003.11.087

Serrano-Notivoli R, Beguería S, Saz MÁ, Longares LA, de Luis M (2017) SPREAD: a high-resolution daily gridded precipitation dataset for Spain—an extreme events frequency and intensity overview. Earth Syst. Sci. Data 9:721–738

Sidorchuk A, Golosov VN (1996) Calibration of soil-erosion models based on study of radioactive fallout. Eurasian Soil Sci 28(10):383–395

Sogon S, Penven M-J, Bonte P, Muxart T (1999) Estimation of sediment yield and soil loss using suspended sediment load and measurements on agricultural land, Brie Plateau, France. Hydrobiol 410:251–261

Soil Survey Staff (1999) Soil taxonomy: a basic system of soil classification for making and interpreting soil surveys, 2nd edn. Agricultural Handbook 436, Natural Resources Conservation Service, USDA, Washington DC, USA, p 869

Soto J, Navas A (2004) A model of ^{137}Cs activity profile for soil erosion studies in uncultivated soils of Mediterranean environments. J Arid Environ 59:719–730. https://doi.org/10.1016/j.jaridenv.2004.02.003

Soto J, Navas A (2008) A simple model of Cs-137 profile to estimate soil redistribution in cultivated stony soils. Radiat Meas 43:1285–1293. https://doi.org/10.1016/j.radmeas.2008.02.024

Spomer RG, McHenry JR, Piest RF (1985) Sediment movement and deposition using cesium-137 tracer. Trans Am Soc Agri Eng 28:767–772

Srisuksawad K, Porntepkasemsan B, Noipow N, Omanee A, Wiriyakitnateekul W, Chouybudha R, Srimawong P (2015) Spatial distribution of ^{137}Cs in surface soil under different land uses in Chao Phraya watershed: potential used as sediment source tracing. J Phys: Conf Ser 611:012026

Suhartini N (2006) Perbandingan Profil Distribusi Vertikal ^{137}Cs Di Lapisan Tanah Hasil Pengukuran Terhadap Simulasi. Makara J Sci 10:89–95

Sutherland RA (1996) Caesium-137 soil sampling and inventory variability in reference samples; literature survey. Hydrol Proc 10:43–53

Takenaka C, Onda Y, Hamajima Y (1998) Distribution of cesium-137 in Japanese forest soils: Correlation with the contents of organic carbon. Sci Total Environ 222:193–199

Tiessen KHD, Li S, Lobb DA, Mehuys GR, Rees HW, Chow TL (2009) Using repeated measurements of ^{137}Cs and modelling to identify spatial patterns of tillage and water erosion within potato production in Atlantic Canada. Geoderma 153(1–2):104–118

Wakiyama Y, Onda Y, Mizugaki S, Asai H, Hiramatsu S (2010) Soil erosion rates on forested mountain hillslopes estimated using ^{137}Cs and $^{210}Pb_{ex}$. Geoderma 159(1–2):39–52

Walling DE, He Q (1999) Improved models for estimating soil erosion rates from cesium-137 measurements. J Environ Qual 28:611–622. https://doi.org/10.2134/jeq1999.00472425002800020027x

Walling DE, Quine TA (1991) Use of ^{137}Cs measurements to investigate soil erosion on arable fields in the UK: potential applications and limitations. J Soil Sci 42:147–165

Walling DE, Quine TA (1992) The use of caesium-137 measurements in soil erosion surveys. Erosion and sediment transport monitoring programmes in river basins. IAHS Publication, Wallingford, pp 143–152

Walling DE, Quine TA (1995) The use of fallout radionuclides in soil erosion investigations. In: Nuclear techniques in soil-plant studies for sustainable agriculture and environmental preservation. IAEA Publication ST1/PUB/947. 597-619

Walling DE, Bradley SB, Wilkinson CJ (1986) A Caesium-137 budget approach to the investigation of sediment delivery from a small agricultural drainage basin in Devon, UK. Proc. symposium on 'Drainage basin sediment delivery.' Albuquerque, IAHS-AISH Publication No. 159:423–435

Wang X, Xue B, Yao S, Yang H, Gu Z, Yang B, Zhang M, Zhu Y (2019) Cs-137 estimates of soil erosion rates in a small catchment on a channelized river floodplain in the lower reaches of Yangtze River. China. J Environ Radioact 208–209:106008. https://doi.org/10.1016/j.jenvrad.2019.106008

Wischmeier WH, Smith DD (1978) Predicting rainfall erosion losses—a guide to conservation planning. U.S. Dept. of Agric., Agriculture Handbook No. 537, U.S. Govt. Printing Office: Washington, DC

Zebarth BJ, Rees HW, Walsh J, Chow TL, Pennock DJ (2002) Soil variation within a hummocky podzolic landscape under intensive potato production. Geoderma 110:19–33

Zhang X, Li S, Wang C, Tan W, Zhao Q, Zhang Y, Yan M, Liu Y, Jiang J (1988) An initial approach to measure soil loss amount of farmland on weir and mound using ^{137}Cs method. Bull Soil Water Conserv 8(5):18–22+29. (in Chinese)

Zhang X, Li SL, Wang C, Tan W, Zhao Q, Zhang Y, Yan M, Liu Y, Jiang J, Xiao J, Zhou J (1989) A study of sediment source from a small catchment in the Loess Plateau by the ^{137}Cs method. Chin Sci Bull 3:210–213 (in Chinese)

Zhang X, Higgit DL, Walling DE (1990a) A preliminary assessment of the potential for using caesium-137 to estimate rates of soil erosion in the Loess Plateau of China. Hydrol Sci J 35:267–276

Zhang X, Li SL, Wang C, Tan W, Zhao Q, Zhang Y, Yan M, Liu Y, Jiang J, Xiao J, Zhou J (1990b) A study of sediment delivery from a small catchment in the Loess Plateau by the ^{137}Cs method. Chin Sci Bull 35(1):37–42

Zupanc V, Mabit L (2010) Nuclear techniques support to assess erosion and sedimentation processes: preliminary results of the use of ^{137}Cs as soil tracer in Slovenia. Dela 33:21–36

Chapter 3
Validating ^{137}Cs Resampling Approach by Comparing with Conventional Erosion Plot Measurements: An Example of Cultivated Site in Italy

P. Porto and E. Fulajtar

Abstract This chapter presents a case study from Italy providing the validation of ^{137}Cs method versus the data from erosion plots. This is a unique study from site where the long-term erosion rates were measured at erosion plots and also ^{137}Cs sampling was done two times. It is the only one such study site worldwide. Therefore, it is presented here as a separate chapter. It should be understood as an example how should the ^{137}Cs resampling study be designed and how the ^{137}Cs method should be validated versus the data on soil erosion rates obtained by conventional erosion measurements. This resampling study was done in seven major methodological steps which were described in detail, each step separately. This is an example of optimal approach how to apply the ^{137}Cs resampling method. The resulting data are interpreted by both key resampling approaches (SRA and TRA). However, one such study cannot be sufficient for validation of ^{137}Cs resampling method. It rather should provide a guidance for other authors on the optimal direction of further research in future.

P. Porto (✉)
Dipartimento di Agraria, University Mediterranea of Reggio Calabria, 89122 Reggio Calabria, Italy
e-mail: paolo.porto@unirc.it

Faculty of Geographical Sciences, Kazimierz Wielki University, Bydgoszcz, Poland

CNR. Istituto per i Sistemi Agrari e Forestali per il Mediterraneo, Sezione Ecologia e Idrologia Forestale, Rende, Italy

E. Fulajtar
Joint FAO/IAEA Centre of Nuclear Techniques in Food and Agriculture, International Atomic Energy Agency, Vienna, Austria
e-mail: emil.fulajtar@nppc.sk

Soil Science and Conservation Research Institute, National Agricultural and Food Centre, Bratislava, Slovakia

P. Porto et al. (eds.), *Using ^{137}Cs Resampling Method to Estimate Mean Soil Erosion Rates for Selected Time Windows*, Environmental Science and Engineering,
https://doi.org/10.1007/978-3-031-52807-1_3

3.1 Introduction

The choice of experimental area in Southern Italy offers a great advantage to test the potential and the strength of ^{137}Cs resampling approach because direct measurements of soil erosion for the period 2006–2016 are available at this site. To the authors knowledge, the only case study available in the literature where a similar experiment could be conducted is based on the dataset provided by Kachanoski (1987) in Canada (see Chapter 2). However, even if the work of Kachanoski is related to a period of similar length (11 years) and involves the data of 10 experimental plots, the different land use and crop management characterizing the study area may have introduced additional uncertainty in the measurements. Also, despite the problem of spatial variability was considered by Kachanoski, the ^{137}Cs measurements were carried out using bulk composite samples collected in different points of each plot. This sampling approach responds to the problem of micro-topography impact on the ^{137}Cs inventory but does not provide any information on the ^{137}Cs depth distribution and the evolution of the ^{137}Cs activity along the soil profile over time.

In contrary, the data provided by this study represent a bare plot, on which the vegetation cover is absent since the beginning of the experiment and only a single up and down slope tillage operation was employed during the study period. Also, the details on ^{137}Cs distribution over the soil profile are provided by several sectioned cores collected in different sampling points. This sampling strategy, together with the relative constant conditions associated with soil type and macro-topography, made it possible to control the main external factors affecting soil erosion and, as such, this study should be seen as a unique example for validating the ^{137}Cs resampling approach.

In Italy, the soil erosion rate assessment using FRNs began already in early '80s (Bazzoffi and Panicucci 1983). Later, various aspects of soil erosion processes were studied, for example the contribution of rill and inter rill erosion (Porto et al. 2014b), sediment budget at watershed level (Porto et al. 2004, 2011) and also the resampling method was tested (Porto et al. 2001, 2014a, b, 2018). The gained experience made an opportunity to learn from the past and a case study was designed at study site with erosion plots, and the results of long-lasting erosion plot measurements were used to validate the ^{137}Cs resampling method. For this purpose, a new approach of ^{137}Cs resampling was developed, which would enable to overcome the difficulties and shortcomings of earlier resampling techniques. This new approach comprises of the following 7 steps:

1. The study area was selected. The possibilities to select the study area were limited because there was needed a site where ^{137}Cs sampling was already done, and this sampling had to be done sufficiently long time ago allowing erosion to remove significant amount of soil, which would be well detectable. For this study, a site located in Southern Italy was selected where experimental erosion plots (5 m $\times$ 25 m) are established and where the ^{137}Cs sampling was already done in 2006. The selected study site was characterized for key climate, relief, soil and land use parameters.

2. The first ^{137}Cs sampling campaign undertaken in 2006 was recalled (Porto and Walling 2012a, b) and the data were scrutinized and evaluated to apply the single sampling (traditional) approach for the time window 1954–2006 and the resampling approach for a second time window. More specifically, the evaluation would have required the ^{137}Cs decay of the old measurements (if only the ^{137}Cs values at the time of sampling would be available) or a reassay of the soil samples for new ^{137}Cs analysis (if original samples would be available). For this study, because the samples are still available at the laboratory of the University Mediterranea, a decision to reanalyse them was made to avoid further uncertainty related to decay calculations and to the fact that the old measurements were made in a different laboratory.
3. The selected study site was resampled in 2017 and the need to relocate the old sampling points was discussed. The data from the new sampling were evaluated and processed to calculate erosion rates for the whole study period (1954–2017) as well as for the second time window (2006–2017).
4. The radiometric analyses were done and gained data were evaluated. The use of earlier reference values or the need for new reference sampling were discussed. The analysis showed that there is no substantial difference between the new reference value obtained in 2017 and the corresponding decay corrected value from the sampling in 2006.
5. Problems due to the different locations of the samples from first and second sampling campaign were identified, compared and it was evaluated how the spatial microvariability can affect the difference between the values from the two compared time periods.
6. Conversion model was selected to process the ^{137}Cs data. A simple proportional model (de Jong et al. 1983) was applied to convert the ^{137}Cs inventories into soil erosion rates. The results of the new sampling done in 2017 were integrated with the values from the first sampling done in 2006.
7. The erosion rates for the second time window (2006–2017) were calculated from the values obtained for first sampling campaign (representing first time window of 1954–2006) and the second sampling campaign (representing the whole study period of 1954–2017). Comparing the values obtained for the first and second time window indicate the changing trend of soil erosion magnitude between the two investigated time windows.

3.2 Study Area Selection (Step 1)

The application of ^{137}Cs resampling approach required the choice of an experimental area that was already investigated for ^{137}Cs analysis. Such study site is available in Southern Italy. It is Caratozzolo Farm in Calabria (38° 16′ N, 15° 49′ E, at an elevation of 585 m ASL, Fig. 3.1). The climate is typically Mediterranean, with a rainy season extending from October to March and a dry summer during which occasional thunderstorms may occur. The mean annual precipitation (Arpacal datasets 1939–2019), obtained at the rainfall station of Santa Cristina d'Aspromonte, is ca. 1470 mm. Mean annual temperatures range from 14 °C to 15 °C, based on the local reports. The land use is characterized by cultivated soils with crop rotation traditional for Calabria (wheat, rye, oat, sorghum, potato, aubergine, cauliflower, horseradish, lupin, tall fescue, and trifolium constituting variable crop rotations). In 2005, the Department of Agraria of the University 'Mediterranea' of Reggio Calabria established five (5 m × 25 m) experimental plots (Fig. 3.1) to explore the effect of different cropping systems on runoff and soil loss (Preiti et al. 2017, 2022). The plots were built on a 10% slope occupied by Typic Hapludands characterised by a silty loam texture (20% sand, 76% silt, and 4% clay). The experimental setup included two levels of disturbance and ground cover (conventional and conservation tillage) combined with several crops in rotation. A bare reference plot, according to Wischmeier and Smith (1978), was also established, as a control system for evaluating the effect of cultivation and the calculation of the C-factor (for soil and crop management details see Preiti et al. 2017). Runoff and soil loss were measured for each plot, at event scale, for the 2006–2016 period (Porto et al. 2022).

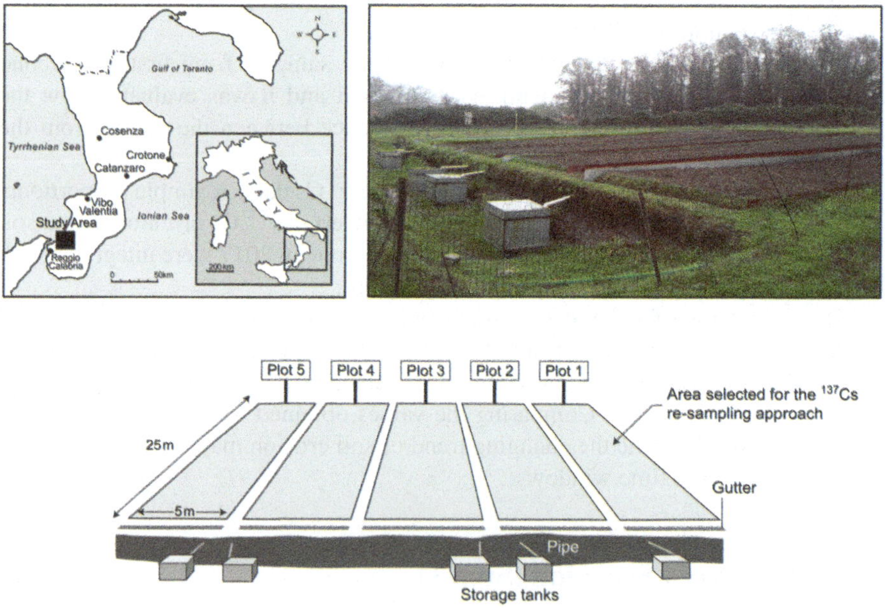

Fig. 3.1 Study area and the experimental plots

3.2.1 *Conventional Measurements of Soil Loss at Experimental Plots*

The measurements of soil loss used in this experiment carried at Plot 1 (bare soil) for which 135 rainfall events that have occurred during the period from January 2006 to December 2016 are available. More specifically, the runoff and soil loss were collected by gutters installed along the lower end of the plot and diverted into a sized tank located at the base of the plot (see Fig. 3.1). Sampling of water and sediment stored in the tank was undertaken two–three days after the end of each rainfall event. Suspended sediment samples were collected from the tanks and transported to the laboratory of the Department of Agraria where they were oven dried and weighed to determine the sediment concentration. The soil loss of each event is calculated as a product of the mean sediment concentration and the total runoff volume collected in the tank during the recorded event. The annual soil loss is in turn calculated as the sum of soil loss of all rain events occurring during the year.

3.3 First and Second Sampling Campaign for ^{137}Cs Measurements (Steps 2 and 3)

Soil sampling for ^{137}Cs analysis was carried out in two different campaigns. The first sampling campaign (*Step 2*) was undertaken in December 2006, and it was designed to establish the ^{137}Cs inventory of the erosion plot. To minimize the disturbance of the plot, only one sectioned core was taken using a steel core tube (diameter of 11 cm). The core tube was inserted into the ground by a motorised percussion corer to a depth of ca. 50 cm. The resulting core was sectioned into 2–4 cm depth increments to explore the ^{137}Cs activity along the soil profile (Porto and Walling 2012a, b). Information about the ^{137}Cs reference value for the first campaign was also available and it was derived from an adjacent cultivated area with no evidence of erosion or deposition processes. The ^{137}Cs reference value and the ^{137}Cs inventory of the plot allowed the application of the ^{137}Cs traditional technique to obtain the erosion rate of the 1954-2006 time window.

The second sampling campaign (*Step 3*) was undertaken in December 2017 using the same equipment. More specifically, the campaign was designed to assess the ^{137}Cs distribution within the study plot and to provide details on the ^{137}Cs reference value from an undisturbed area. In this case, eight sectioned cores (with the same depth increments) were collected from the study plot to check the micro-scale variability of the ^{137}Cs inventory. A composite sectioned core was obtained from the reference area (the same cultivated location used for the first sampling campaign). The new ^{137}Cs reference value and the ^{137}Cs inventory values of the erosion plot allowed the application of the ^{137}Cs traditional technique to obtain the erosion rate for the whole study period (1954–2017).

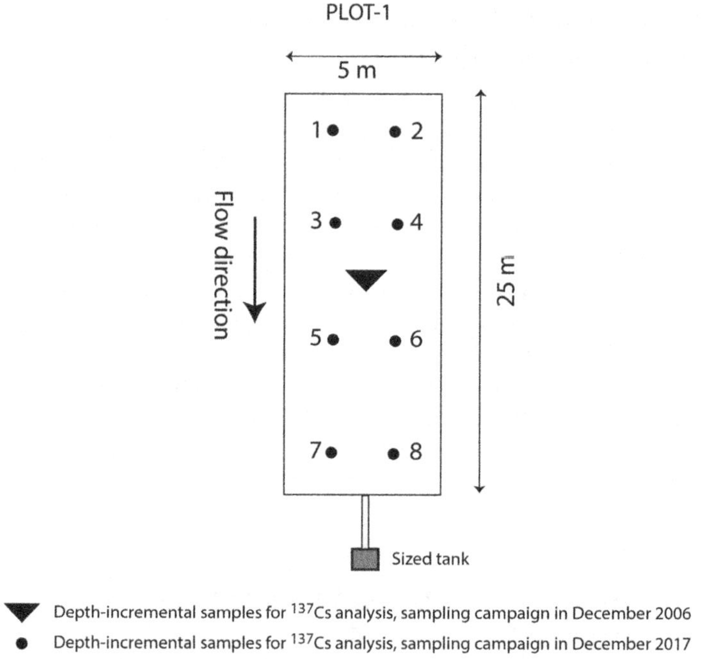

Fig. 3.2 Location of the soil samples collected in Plot 1 during the sampling campaigns in 2006 and 2017

The location of each sample for the two sampling campaigns is illustrated on Fig. 3.2.

3.4 Data Analysis and Evaluation (Step 4)

3.4.1 Radiometric Analyses

The soil samples collected in 2006 were archived and are still available at the University of Reggio Calabria and in 2017 they were again processed for radiometric analyses using the same protocol as was used for new samples collected in 2017. Prior to the analyses of ^{137}Cs activity, the samples were oven dried at 45 °C and weighed to calculate the mass depth. After disaggregating and sieving to <2 mm, each sample was mixed for 2 h using a 3D shaker to homogenize the ^{137}Cs content. Then, representative sub-samples were placed into Petri dishes or Marinelli beakers of different sizes depending on the sample mass. The samples were measured by two Canberra p-type high-resolution low energy coaxial HPGe detectors (model GX4020) with a relative efficiency of 45%. The detectors are calibrated using the software LABSOCS (Laboratory Sourceless Calibration Software) and a multigamma source. Counting time varied from ca. 80,000 to ca. 240,000 s depending on the activity that was expected for each sample. All the samples were analysed (or reanalysed) in 2018 at the University of Reggio Calabria and the resulting measurements have been standardised to a fixed date at the end of 2017.

3.4.2 Depth Distribution of ^{137}Cs Inventories

The ^{137}Cs inventories determined from the sectioned cores collected from the plot and the reference area in 2006 and 2017 are reported in Table 3.1. The location of each profile is indicated in Fig. 3.2.

It is clear from Table 3.1 that, even if the plot was maintained bare under constant tillage operations and, as such, a uniform ^{137}Cs content should be expected over the area, some variability in ^{137}Cs inventory, mainly related to micro-topography (due to possible accumulation of rain-splashed soil particles in temporary micro-depressions and bioturbation) can be observed. This is demonstrated by the 9 corresponding ^{137}Cs profiles, 1 for the 2006 sampling campaign and 8 for the 2017 sampling campaign, illustrated in Fig. 3.3.

However, the information reported in Table 3.1 and Fig. 3.3 indicates that all the inventories obtained from the resampling in 2017 are lower than the inventory value (2197 Bq m^{-2}) obtained from the single sectioned core collected in 2006 (and decayed to 2017). This result suggests that, during the study period 2006–2017, soil redistribution occurred within the plot and erosion was the dominant process.

Another important indication that can be inferred from Fig. 3.3 is related to the ^{137}Cs depth distribution when comparing the eight profiles of the 2017 sampling

Table 3.1 Values of ^{137}Cs inventory obtained for the two sampling campaigns in 2006 and 2017

2006 Sampling campaign	^{137}Cs Inventory (Bq m^{-2})	2017 Resampling campaign	^{137}Cs Inventory (Bq m^{-2})	^{137}Cs reference details	
				Year	Value (Bq m^{-2})
Single profile	2197	Profile 1	1547		
		Profile 2	1811		
		Profile 3	1829		
		Profile 4	1436	2006	5321[a]
		Profile 5	1618	2017	4233
		Profile 6	1706		
		Profile 7	1757		
		Profile 8	998		

The last two columns indicate the reference values related to the first and the second sampling campaign

[a]The reference value for the 2006 campaign was derived from the ^{137}Cs activity measured in 2018 (analysing the soil samples collected in 2006) and decay corrected to 2006

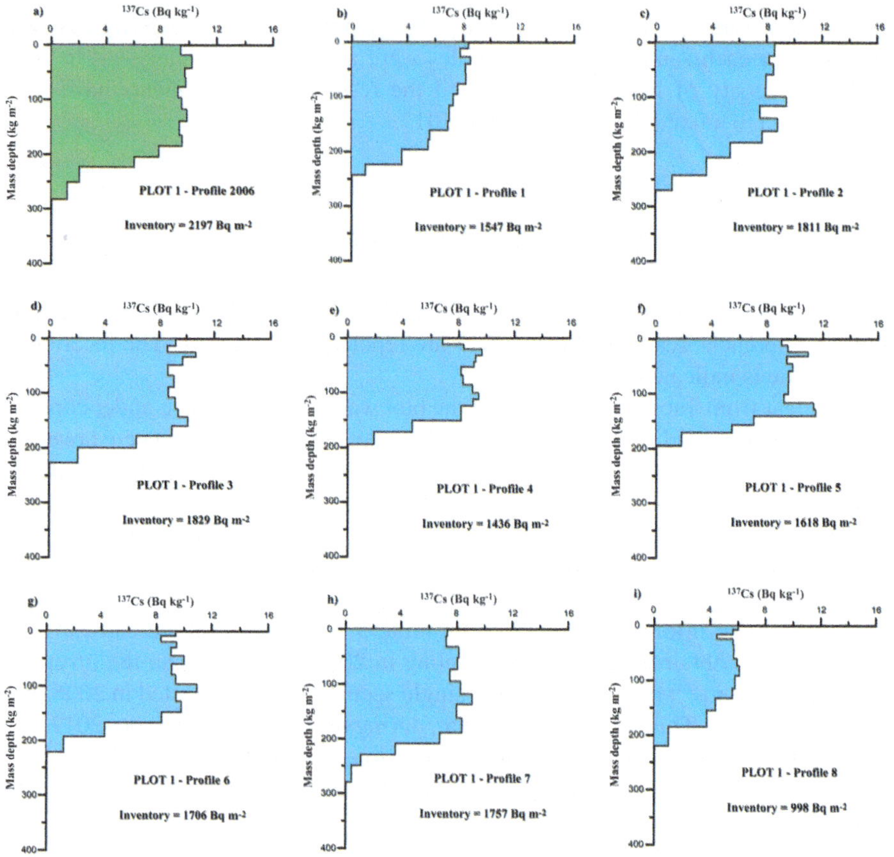

Fig. 3.3. Depth distribution of ^{137}Cs inventories corresponding to the 9 sectioned cores collected in Plot 1: 1 profile sampled in 2006 (a) decay corrected to the end of 2017 and 8 profiles sampled in December 2017 (b-i)

campaign (Fig. 3.3b–i) with the profile sampled in 2006 (Fig. 3.3a). The profile illustrated in Fig. 3.3a would reproduce the ^{137}Cs activity of the original 2006 soil profile as if no external disturbance (erosion or deposition) had taken place during the study period. In other words, the application of the resampling approach in this area assumes that the only changes occurred in the 2006 inventory are related to the reduction of the ^{137}Cs content due to decay. From this point of view, the 2006 profile shows evidence of cultivation, because ^{137}Cs is seen to be relatively well mixed within the plough layer (top ca. 15–20 cm), and a decreasing activity is documented for deeper soil layers (down to ca. 30 cm).

The profiles sampled in 2017 (Fig. 3.3b–i) have the ^{137}Cs activity lower in each sampled soil layer than the profile sampled in 2006 (Fig. 3.3a). Also, the thickness of the ^{137}Cs contaminated layer is lower in profiles from 2017 as compared to 2006

(20–25 cm versus 30 cm). These effects reflect the impact of erosion active between 2006 and 2017.

3.4.3 *Choice of ^{137}Cs Reference Inventories: Earlier Values or New Sampling?*

Application of the ^{137}Cs traditional technique requires the use of a ^{137}Cs reference value representative for the study area. This value is available for both sampling campaigns of 2006 and 2017 and it was obtained by sampling the same reference area located in the vicinity of the study plot. Details of these values that include the total inventory and the ^{137}Cs distribution over the soil profile are reported in Table 3.1 and in Fig. 3.4a,b, respectively.

The reference value obtained for the first sampling in 2006 was used to obtain the erosion rate for the first time window (1954–2006), while the value obtained for the second sampling in 2017 served to obtain the erosion rate for the whole study period (1954–2017). It is worth to realise that if the ^{137}Cs reference value is available for the first sampling campaign, the second sampling is not strictly necessary because the ^{137}Cs reference value corresponding to time of resampling can be obtained by decay correcting the value from the first sampling. In this particular case, the ^{137}Cs reference value obtained in 2006 (5321 Bq m^{-2}) was decay corrected to the date of the sampling campaign in 2017, and the resulting value of 4134 Bq m^{-2} was very close to that (4233 Bq m^{-2}) obtained by sampling the reference site in 2017. If a reference value from the first sampling is not available, the reference sampling should be done during the second sampling campaign and the obtained value can be decay corrected to the date of the first sampling to get the values of the erosion rate for both periods.

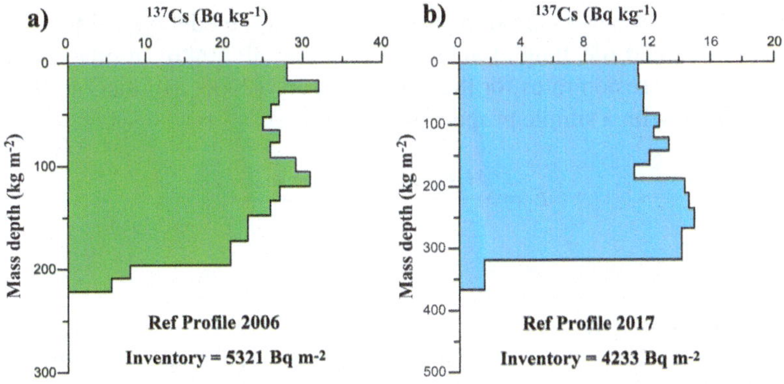

Fig. 3.4. Depth distribution of ^{137}Cs inventories of the reference profiles sampled in December 2006 (a) and December 2017 (b)

3.5 Problems Due to the Different Locations of the Samples Collected During the Two Sampling Campaigns and the Effects of Spatial Micro-Variability (Step 5)

A rigorous application of the resampling approach required a relocation of each sample to compare each ^{137}Cs inventory of the second sampling with the corresponding inventory of the first sampling. Theoretically, considering the small investigated area and the absence of vegetation, this operation would have been very easy but there are practical reasons making this comparison difficult. For example, there is a need to avoid sampling the same point, which was sampled already because it ha s been disturbed by the previous sampling operation (see Porto et al. 2014a). Also, because the inherent small-scale variability of ^{137}Cs inventories within the sampling spots is expected, collection of a new sample at a short distance from the corresponding first sample could provide biased estimates. For these reasons, it would be useful to capture such variability and to avoid biasing by collecting several cores in the same spot. In this case, because only a single sectioned core was available for the 2006 campaign (assuming that it was representative of the whole plot), we decided to collect eight sectioned cores (with the same depth increments) to take into account the micro-space variability associated with the 2006 sampling point.

3.6 Selection of the Conversion Model to Convert the ^{137}Cs Data into Soil Erosion Rates (Step 6)

As indicated above, the plot area was maintained bare for the duration of the experiment (11 years). Also, the area was subjected to cultivation since the early 1950s and the ^{137}Cs fallout inputs can be considered completely mixed within the cultivated layer (ca. 0.25–0.30 m) as it is documented by the profiles on Fig. 3.3. In this case, if the total ^{137}Cs fallout occurred in 1963, a conversion model typical for cultivated soils can be applied (de Jong et al. 1983; Zhang et al. 1990). In other words, the estimates of soil erosion rates for the period of 1963–2006, and that of 1963–2017 can be obtained with a simple proportional model:

$$E = \frac{M_s(Cs_{ref} - Cs_f) * 10}{(t_i - t_f)Cs_{ref}} \tag{3.1}$$

where:

E is erosion rate (t ha^{-1} yr^{-1});
M_s is specific mass of the plough layer (kg m^{-2}), calculated from the average bulk density (1000 kg m^{-3}) and the average depth of the plough layer (0.25 m);

t_i is time of ^{137}Cs fallout in 1963 (yr); the conversion models use the year 1963 instead of 1954 because 1963 was the year with maximum ^{137}Cs fallout (several-fold greater than in 1954).
t_f is time of the ^{137}Cs sampling (yr);
Cs_{ref} is total ^{137}Cs reference inventory (Bq m^{-2}) (see Table 3.1 for details);
and Cs_f is total ^{137}Cs inventory related to the ^{137}Cs sampling campaign (Bq m^{-2}) (see Table 3.1 for details).

The choice of proportional model expressed by the Eq. 3.1 is justified by its simplicity, based on the assumption that a linear proportionality between the soil loss and the amount of ^{137}Cs removed from the soil profile exists. Even if this model does not take into account the dilution of ^{137}Cs concentrations in the soil within the plough layer (see the above considerations), and it is based on the assumption that the total ^{137}Cs fallout occurred in 1963, we are confident that these effects are very low and can be neglected for this case study.

For these reasons, Eq. 3.1 was used to obtain the value of erosion rate corresponding to the single sampling point sampled in 2006 and the values of erosion rate corresponding to the eight sampling points sampled in 2017. These values are reported in Table 3.2.

Table 3.2 indicates the soil loss of 26.8 t ha^{-1} yr^{-1} for the 1963–2006 period, and for the 1963–2017 period (derived from the eight sampling points sampled in 2017) it ranges from 25.8 t ha^{-1} yr^{-1} to 35.1 t ha^{-1} yr^{-1} and the mean value of 28.5 t ha^{-1} yr^{-1} is greater than that of 26.8 t ha^{-1} yr^{-1} suggesting that erosion rate increased during the second time window.

Table 3.2 Values of erosion rate obtained for the periods of 1963–2006 and 1963–2017

2006 Sampling	Erosion rate (1963–2006) (t ha^{-1} yr^{-1})	2017 Sampling	Erosion rate (1963–2017) (t ha^{-1} yr^{-1})
Single profile	26.8	Profile 1	29.0
		Profile 2	26.0
		Profile 3	25.8
		Profile 4	30.2
		Profile 5	28.2
		Profile 6	27.2
		Profile 7	26.6
		Profile 8	35.1
		Mean value	28.5

3.7 Results of Conventional Measurements of Soil Erosion Rates at the Experimental Plot

The results of the 11 years lasting erosion monitoring (2006–2016) at the experimental plot are provided on Fig. 3.5. Such data, even if not mandatory for the application of the resampling technique, should be highly recommended because they allow to obtain an independent validation of the used resampling approach.

Figure 3.5 indicates that the measured values of soil loss show a great interannual variability ranging from a minimum of 4.7 t ha^{-1} (2014) to a maximum of 158 t ha^{-1} (2015), which is related to the temporal rainfall distribution. The mean value is ca. 69.0 t ha^{-1}, and the standard deviation (SD) is 44.8 t ha^{-1}. It is worth noticing that these values of soil erosion are very high because they are strictly related to the period of monitoring when the vegetation was absent at the erosion plot and cannot be considered representative of the previous period during which the area was covered by various crops. For this reason, they cannot be directly compared with the estimates of erosion rate provided by the ^{137}Cs traditional technique because the latter covers a longer period that includes a time window of 1963-2006 during which soil erosion was lower due to the presence of vegetation. Consequently, only the application of a resampling approach will make such comparison possible.

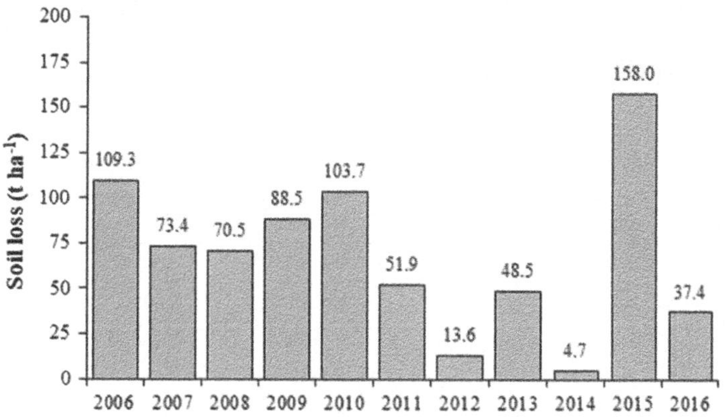

Fig. 3.5 Annual soil loss obtained from the experimental plot during the 2006–2016 period

3.8 Comparing Erosion Rates Obtained by Conventional Measurements at Erosion Plots and by the ^{137}Cs Resampling Technique: A Validation Exercise (Step 7)

The aim of this erosion plot experiment was to compare the mean annual soil loss for the 2006–2017 period, obtained by the ^{137}Cs resampling approach, with the equivalent values of soil loss measured on the erosion plot. This comparison is important as it tests the potential of the ^{137}Cs resampling approach to provide suitable estimates of erosion rate for selected periods and represents an independent validation of the ^{137}Cs resampling method.

The availability of a long time (11 years) data series of soil loss measurements is a unique opportunity to do this validation exercise for the following reasons:

1. The plot was maintained bare, with frequent tillage operations, for the entire duration of the experiment and no effect associated with possible land use changes can be accounted for during the whole 11 years of measurement period.
2. The soil type is the same over the whole plot area and no variation due to different values of soil erodibility can be assumed.
3. No diverse effect of land use and management should be expected because only the up and down slope tillage was employed for the whole measurement period.
4. Finally, although the effect of macro-topography should be considered negligible because the slope is uniform and constant (10%) there is some variability over the plot surface that can be attributed to the presence of microrelief such as minor slope undulations and the temporary surface roughness resulting from the tillage operations.

However, when comparing the soil erosion rates based on ^{137}Cs measurements, with the direct soil loss measured at erosion plots, the problem due to the spatial variability of the ^{137}Cs content, associated with the micro-topography, must be considered. The interannual variation of soil loss due to the rainfall temporal distribution must be considered as well.

The first problem is addressed by the availability of eight ^{137}Cs profiles for the 2017 campaign. This sampling strategy offers the opportunity to evaluate the effect of micro-topography as the sampling points are very close to each other (only a few meters distant) and the corresponding estimates of soil erosion rate can be assumed representative for the plot area. We recognize that an ideal application of the resampling method and a more rigorous uncertainty analysis would have required collection of the same number of cores in both campaigns. Unfortunately, this was not possible because the sampling campaign in 2006 was limited by the need to avoid soil disturbance at that time and only one sectioned core was collected. However, the 2006 sampling point is located in the central part of the plot and the corresponding inventory is considered representative for the whole plot.

The second problem is that the temporal variability of the annual rainfall distribution obviously causes the temporal variability of soil erosion. However, considering the short time window between the two sampling campaigns, this effect is well explained by the direct measurements of soil loss that cover a large range of values (Fig. 3.5).

For the above reasons, the availability of two sets of soil erosion rates suggests making this comparison by considering their range instead of their mean to account for their uncertainty.

The available ^{137}Cs datasets and the corresponding erosion rates for the two periods of 1963–2006 and 1963–2017 make possible to apply the two basic resampling approaches (see Chapter 1), namely the traditional reference approach (TRA) and spatial reference approach (SRA), to get the soil erosion rates for the 2006–2017 period. These estimates can be compared with the direct observations of soil loss obtained during the 11-year monitoring of the erosion plot.

3.8.1 *Spatial Reference Approach*

This procedure does not require the ^{137}Cs reference value obtained from the stable area located in the vicinity of study plot but it assumes that the ^{137}Cs inventory (2197 Bq m^{-2}) obtained within the plot during the first sampling campaign in 2006 (Table 3.1) will serve as the reference value for the calculations. Based on this assumption, the proportional model expressed by the Eq. (3.1) provides the eight values of soil erosion rate listed in Table 3.3 that range from a minimum of 38.0 t ha^{-1} yr^{-1} to a maximum of 124.0 t ha^{-1} yr^{-1}, with a mean of 63.0 t ha^{-1} yr^{-1} and a SD of 28.3 t ha^{-1} yr^{-1}.

Table 3.3 The estimates of soil erosion rate provided by the two resampling approaches for the 2006–2017 period

Sampling point	Erosion rate (t ha^{-1} yr^{-1})	
	Spatial reference approach	Traditional reference approach
Profile 1	67.3	35.7
Profile 2	40.0	21.2
Profile 3	38.0	20.2
Profile 4	78.8	41.9
Profile 5	59.9	31.9
Profile 6	50.7	27.0
Profile 7	45.6	24.2
Profile 8	124.0	65.9
Mean Value	63.0	33.5

3.8.2 Traditional Reference Approach

This procedure requires the information reported in Table 3.2, provided by application of the traditional ^{137}Cs technique for 1963–2006 and 1963–2017 time windows. The mean value of erosion rate obtained for 1963–2006 time window (26.8 t ha^{-1} yr^{-1}) and 1963–2017 time window (28.5 t ha^{-1} yr^{-1}) needs to be integrated for the entire period covered by the ^{137}Cs measurements and then divided by the number of years elapsed since the two sampling campaigns (11 years). This procedure provided the estimates of soil erosion rate reported in Table 3.3 together with those obtained using the SRA. More specifically, the values range from a minimum of 20.2 t ha^{-1} yr^{-1} to a maximum of 65.9 t ha^{-1} yr^{-1}, with a mean value of 33.5 t ha^{-1} yr^{-1} and a SD of 15.0 t ha^{-1} yr^{-1}.

To facilitate interpretation of the results listed in Table 3.3, these ranges are reported in graphical form (see horizontal bars) on Fig. 3.6 together with the range of the measured values of soil loss. The comparison on Fig. 3.6 shows that the ranges of soil loss estimated with both the SRA and TRA lie within the range of soil loss related to the direct measurements. This is a first successful validation of the ^{137}Cs resampling method.

However, it should be recognised that the range of the plot measurements is larger than that of the estimates provided by the ^{137}Cs measurements as it contains two outliers associated with the lowest value recorded in 2014 and the highest value recorded in 2015 (see Fig. 3.5). If the comparison is restricted to the mean values, and their relative uncertainty is evaluated according to their respective standard error, it is clear that the TRA tends to underestimate the rate of soil erosion for the period covered by the measurements on erosion plot.

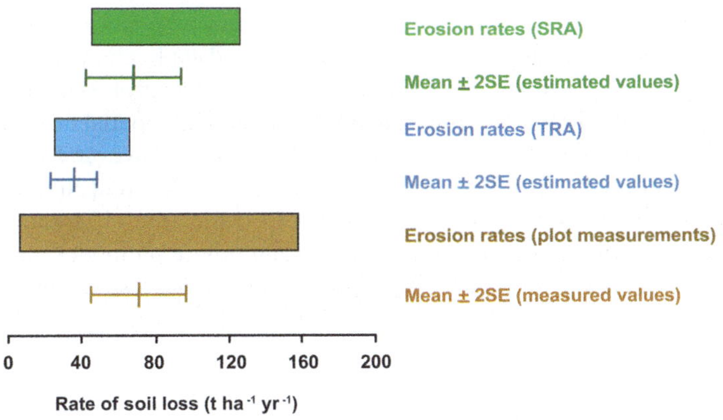

Fig. 3.6 Comparing the ranges of mean annual soil loss obtained by the SRA and TRA for the 2006–2017 period and the soil loss measured on the experimental plot for the 2006–2016 period; The uncertainty of the data is expressed by the mean values ± 2SE

On the contrary, the results provided by the SRA are very encouraging as the erosion rate obtained by this approach and represented as the mean value ± 2SE on Fig. 3.6 is almost overlapping with that of erosion plot measurements.

This result can be considered as very important test of validation because it suggests that despite of the uncertainty associated with the ^{137}Cs measurements, the SRA, combined with the use of a simple proportional model, is a very reliable method to provide estimates of soil loss for time windows shorter than those usually employed for the ^{137}Cs traditional approach. At the same time, it is difficult to explain the underestimation provided by the TRA. A possible explanation can be related to the uncertainty involved in the calculation of the reference value that affected the estimates related to both sampling campaigns. Also, the sudden change of land use occurred in 2006 (since when the plot was maintained bare) may have introduced an additional uncertainty in the calculation of the mean values of erosion rate provided by ^{137}Cs and this uncertainty may have propagated over the study period causing a biased value of soil erosion rate. However, it is worth noticing that these results are in line with those obtained by Fornes et al. (2005) and those derived from the results of Kachanoski (1987). In both cases, the TRA did not provide satisfactory agreement with the measured data showing, respectively, underestimation and overestimation of soil erosion rates. On the contrary, the SRA provided estimates of soil erosion very close to the erosion plot measurements.

3.9 Conclusions

The results obtained in this case study represent a first attempt to test the ability of the ^{137}Cs resampling approach at cultivated land. The close agreement between the soil erosion rate provided by the SRA and the soil loss measurements at erosion plots is very encouraging for the future and suggests an extended application of the method in different geomorphic contexts. The combined use of a simple proportional model and the selection of several sampling points to take account of the spatial microvariability proved to be a very effective strategy to apply this method. However, further work is necessary to confirm these results in different locations and different climatic and land use environments, including uncultivated areas for which the different conversion models (depth distribution model and diffusion and migration model) are used and where the problems of spatial and temporal variability of soil erosion rates are greater.

References

Bazzoffi P, Panicucci M (1983) Erosione sui versanti e conseguente sedimentazione in piccoli serbatoi artificiali. Annali Ist. Sperimentale Studio e Difesa Suolo, Firenze 14:127–178

de Jong E, Begg CBM, Kachanoski RG (1983) Estimates of soil erosion and deposition for some Saskachewan soils. Can J Soil Sci 63:607–617

Fornes WL, Whiting PJ, Wilson CG, Matisoff G (2005) Caesium-137-derived erosion rates in an agricultural setting: the effects of model assumptions and management practices. Earth Surf Process Land 30:1181–1189

Kachanoski RG (1987) Comparison of measured soil 137-Cesium losses and erosion rates. Can J Soil Sci 67:199–203

Porto P, Walling DE, Callegari G (2004) Validating the use of caesium-137 measurements to estimate erosion rates in three small catchments in Southern Italy. IAHS Publication, 288:75–83

Porto P, Walling DE, Callegari G (2011) Using ^{137}Cs measurements to establish catchment sediment budgets and explore scale effects Abstract Hydrological Processes 25(6) 886–900 https://doi.org/10.1002/hyp.v25.6 10.1002/hyp.7874

Porto P, Walling DE (2012a) Validating the use of ^{137}Cs and $^{210}Pb_{ex}$ measurements to estimate rates of soil loss from cultivated land in southern Italy. J Environ Radioact 106:47–57

Porto P, Walling DE (2012b) Using plot experiments to test the validity of mass balance models employed to estimate soil redistribution rates from ^{137}Cs and $^{210}Pb_{ex}$ measurements. Appl Radiat Isot 70:2451–2459

Porto P, Walling DE, Ferro V (2001) Validating the use of caesium-137 measurements to estimate soil erosion rates in a small drainage basin in Calabria, Southern Italy. J Hydrol 248(1–4): 93–108

Porto P, Walling DE, Alewell C, Callegari G, Mabit L, Mallimo N, Meusburger K, Zehringer M (2014a) Use of a ^{137}Cs re-sampling technique to investigate temporal changes in soil erosion and sediment mobilisation for a small, forested catchment in southern Italy. J Environ Radioact 138:137–148

Porto P, Walling DE, Capra A (2014b) Using ^{137}Cs and $^{210}Pb_{ex}$ measurements and conventional surveys to investigate the relative contributions of inter rill/rill and gully erosion to soil loss from a small, cultivated catchment in Sicily. Soil Tillage Res 135:18–27

Porto P, Bacchi M, Preiti G, Romeo M, Monti M (2022) Combining plot measurements and a calibrated RUSLE model to investigate recent changes in soil erosion in upland areas in Southern Italy Journal of Soils and Sediments 22(3) 1010–1022 https://doi.org/10.1007/s11368-021-03119-2

Preiti G, Romeo M, Bacchi M, Monti M (2017) Soil loss measure from Mediterranean arable cropping systems: effects of rotation and tillage system on C-factor. Soil Tillage Res 170:85–93

Preiti, G, Calvi, A, Porto, P, Romeo, M, Monti, M, Bacchi, M (2022) Long-term effects of different arable cropping systems on surface erosion processes and C-factor in hilly Mediterranean environment Soil and Tillage Research 223105480- https://doi.org/10.1016/j.still.2022.105480

Porto P, Walling DE, Callegari G (2018) Using repeated ^{137}Cs and ^{210}Pbex measurements to establish sediment budgets for different time windows and explore the effect of connectivity on soil erosion rates in a small experimental catchment in Southern Italy. Land Degrad Dev 29:1819–1832

Wischmeier WH, Smith DD (1978) Predicting rainfall erosion losses—a guide to conservation planning. U.S. Dept. of Agric., Agriculture Handbook No. 537. U.S. Govt. Printing Office: Washington, DC. 60p

Zhang XB, Higgitt DL, Walling DE (1990) A preliminary assessment of the potential for using caesium-137 to estimate rates of soil erosion in the Loess Plateau of China. Hydrol Sci J 35:267–276

Chapter 4
Step-By-Step Protocol to Apply the ^{137}Cs Resampling Approach to Assess Soil Erosion During Chosen Time Periods

P. Porto and E. Fulajtar

Abstract This chapter provides a step-by-step protocol to apply the ^{137}Cs resampling approach to assess soil erosion during chosen time periods, which is a final result of these guidelines. It is based on the results of Chapter 3, but the steps presented in that chapter are adjusted for an individual case of the Caratozzolo Study Site presented in Chapter 3. These steps were generalised to form of a universally applicable protocol of ^{137}Cs resampling procedure comprising of 7 steps but with two modifications, one for the traditional reference approach (TRA) and the second for the spatial reference approach (SRA). These steps are following: (1) Selecting and assessing study site and existing ^{137}Cs data, (2) Selecting the methodical approach (TRA versus SRA), (3) Second sampling campaign of selected study site, (4) Sample preparation and gamma spectroscopic analyses, (5) Determining the reference value (recalculating the earlier reference value for radioactive decay or new reference sampling), (6) Conversion of ^{137}Cs inventories from second sampling to soil redistribution rates, (7) Calculating soil redistribution rates for the time window since the first until the second sampling). These steps are characterized in a graphical scheme presenting also the two versions of this approach aimed on TRA and SRA. Finally, this chapter provides a discussion on the further research and proposes some possible objectives of next ^{137}Cs resampling studies.

P. Porto (✉)
Dipartimento di Agraria, University Mediterranea of Reggio Calabria, 89122 Reggio Calabria, Italy
e-mail: paolo.porto@unirc.it

Faculty of Geographical Sciences, Kazimierz Wielki University, Bydgoszcz, Poland

CNR. Istituto per i Sistemi Agrari e Forestali per il Mediterraneo, Sezione Ecologia e Idrologia Forestale, Rende, Italy

E. Fulajtar
Joint FAO/IAEA Centre of Nuclear Techniques in Food and Agriculture, International Atomic Energy Agency, Vienna, Austria
e-mail: emil.fulajtar@nppc.sk

Soil Science and Conservation Research Institute, National Agricultural and Food Centre, Bratislava, Slovakia

P. Porto et al. (eds.), *Using ^{137}Cs Resampling Method to Estimate Mean Soil Erosion Rates for Selected Time Windows*, Environmental Science and Engineering,
https://doi.org/10.1007/978-3-031-52807-1_4

4.1 Protocol for Application of ^{137}Cs Resampling Method

The usual approach of using ^{137}Cs technique based on one sampling campaign has these important limitations:

1. The estimates of soil redistribution rates are time related. This means that, if the traditional approach is employed, the calculated soil redistribution rates represent the mean values for the whole time since the ^{137}Cs fallout (1954) until the time of sampling. This time is still growing and recently it is already too long (ca. 60 years) what is too much for many practical purposes. This makes the technique unsuitable to detect changes in erosion rates during shorter time windows what is important especially for studying the impacts of land use changes and climate changes.
2. There is uncertainty related to the choice of a conversion model able to convert ^{137}Cs redistribution to soil redistribution and the limited availability of input parameters for some models. The conversion models need calibration and validation using local parameters and this limitation is often overlooked or ignored in many studies.
3. In areas affected by Chernobyl or Fukushima fallout it is usually difficult to distinguish the proportion of ^{137}Cs inputs from bomb tests and nuclear power plant accident and thus the time of ^{137}Cs input is uncertain. This increases the overall uncertainty of the ^{137}Cs method in Chernobyl and Fukushima affected areas, and it makes the use of conversion models much more difficult and the results much more uncertain.
4. The spatial heterogeneity (microvariability) of ^{137}Cs inventories and the representativeness of reference site.

On the other hand, the very long period since the ^{137}Cs fallout is very useful and advantageous for some ^{137}Cs applications, especially for studying sediment chronology. In such case, when investigating the overbank sedimentation on floodplains or sedimentation in water reservoirs, lakes and coastal zones of sea the different peaks of radionuclide contamination along the sampled soil or sediment profiles can be attributed to certain events such as two peaks associated with bomb fallout (1958 and 1963), nuclear accidents (Chernobyl, 1986 or Fukushima, 2011) and in some areas local peaks can be detected, which are associated with some significant land use changes or natural disasters known from the land use history. This provides an opportunity to derive time integrated deposition estimates covering different time windows. However, the peaks associated with nuclear accidents, which are very helpful for this approach are limited to relatively small areas affected by the nuclear accidents. Chernobyl fallout affected Europe and nearby Northern Africa and West Asia and Fukushima affects only Japan. Moreover, the distinction between bomb ^{137}Cs and powerplant accident ^{137}Cs is conspicuous only in those areas, where the powerplant ^{137}Cs is dominant, but in areas where the powerplant fallout is of the same order of magnitude as the bomb fallout the quantification of the two sources is difficult.

The above discussed limitations can be overcome by the ^{137}Cs resampling. For that reason, the resampling method was considered already long time ago (since 1980s), but it was used only occasionally, because the resampling studies need more complex and more systematic design and project management. When this method will be sufficiently standardized, it will spread more widely, and will be used routinely., Only then the full potential of ^{137}Cs tracer will be exploited.

This guidelines is a first attempt to provide a comprehensive methodological guidance for better (and to some extend standardised) approach of soil erosion assessment using the ^{137}Cs resampling. Achieving this goal is based on reviewing existing studies using ^{137}Cs resampling (Benmansour et al. 2019; Chappell et al. 2015; de Jong and Kachanoski 1988; Kachanoski and de Jong E 1984; Li et al. 2011; Lobb and Kachanoski 1999; McHenry and Bubenzer 1985; Moustakim et al. 2019; Porto et al. 2018; Tiessen et al. 2009) and on own investigation carried by authors of this guidelines. Own case studies were carried in Italy, Spain, Morocco and China representing a great variability of geographical conditions from moist to dry and from temperate to warm climates. As it was presented in Chapter 1 (Table 1.1) in total 14 publications describing 54 case studies were reviewed. These studies were carried out by a great number of research teams using different research approaches. All information gained from literature as well as from own studies was summarised and the systematic assessing of the existing approaches of the ^{137}Cs resampling case studies was presented. The assessment was aimed especially on comparing all used methodological approaches. Several different approaches were observed in evaluated studies, but all used approaches were based on one of two basic principles. Therefore all evaluated approaches were grouped into two groups according to which of these two identified principles were used.

1. The first approach is called ***Traditional Reference Approach*** (***TRA***). It subtracts the time weighted erosion rate of the period since the ^{137}Cs fallout (1954) until the first sampling, from the erosion rate of the period since the ^{137}Cs fallout (1954) until the second sampling (see Fig. 4.1). This method requires a reference value for each sampling campaign that is obtained in most cases by sampling a reference site (representing undisturbed area) situated close to the study site, although instead of that the direct ^{137}Cs fallout measurements can be used where available. This is, however, rare, and it can be done only in those areas where such measurements were carried and it is rather exceptional, because radionuclide fallout measurements were rare even in technologically most advanced countries. Thus, usually this approach requires reference ^{137}Cs inventory from reference site and this approach is like the traditional single sampling ^{137}Cs technique (hence the term 'traditional reference approach'). The reference site can be sampled two times during both first and second sampling campaign, but usually this is not feasible and in most cases the reference site was sampled only once, and the missing reference value was derived from the available value corrected for the decay to the time of that sampling campaign for which the reference value was not sampled. Usually, the reference sampling use to be done during the first sampling campaign and the reference value for the second sampling campaign

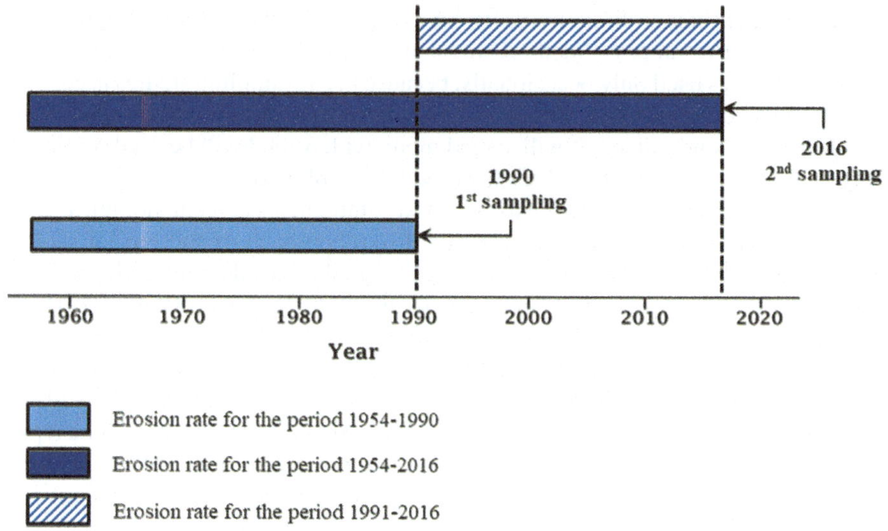

Fig. 4.1 Example of Traditional Reference Approach (TRA) for two sampling campaigns in 1990 and 2016

use to be calculated by decay correction of the reference value obtained from the first sampling campaign. Nevertheless, the opposite approach is possible as well.

2. The second approach, called ***'Spatial Reference Approach' (SRA),*** assumes that the ^{137}Cs inventories measured during the first campaign at the studied site can serve as reference values for the second sampling (Fig. 4.2). In this way each sampled point has own reference value as the ^{137}Cs inventory of each point sampled during the second campaign is compared to that of the corresponding point sampled during the first campaign (decay corrected to the time of the second sampling campaign). In such case each sampling point has its own point reference value instead of one mean value of reference site. Thus, the reference value for the second period is spatially differentiated (hence the term 'spatial reference approach').

The major difference between these two approaches is in the different concept of reference values. The principle of reference value is a core idea of soil erosion assessment using radionuclide tracers. The inventory of ^{137}Cs at a studied site is compared to the reference ^{137}Cs inventory representing the original soil contamination of the study area. The successful application of the ^{137}Cs technique is based on the selection of appropriate reference value and accuracy of the results depends on how precisely the selected reference values represent the original soil contamination. In principle three possibilities to obtain reference value exist:

a. The radionuclide fallout measurements can be used where available;
b. The radionuclide inventory of undisturbed (noneroded) reference sites can be used;

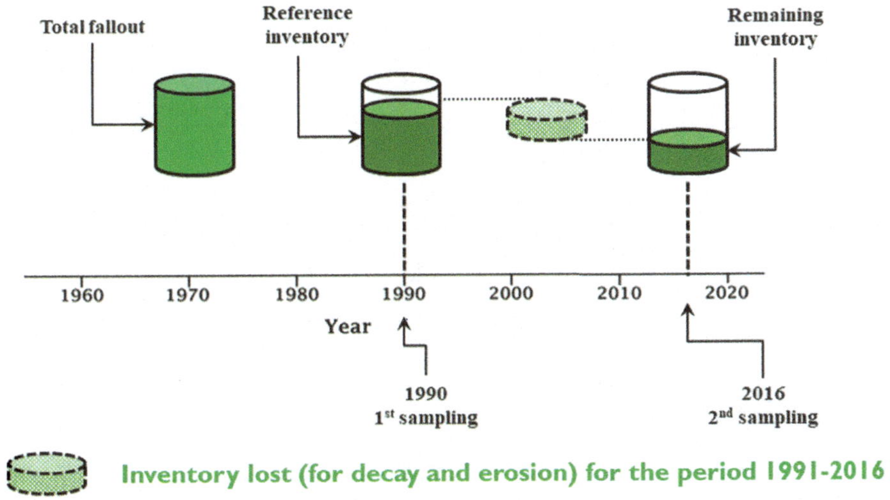

Fig. 4.2 Example of Spatial Reference Approach (SRA) for two sampling campaigns in 1990 and 2016

c. The radionuclide inventories measured directly at the individual sampling points of the studied site during first sampling campaign can be used as a reference value for the corresponding points sampled during the second sampling campaign (resampling strategy).

The third approach is most accurate because it reduces the problems with microvariability of radionuclide spatial distribution and this increases the representativeness of reference site. Therefore, the resampling approach improves the overall accuracy and representativeness of the ^{137}Cs technique.

The information reported in Table 1.1 indicates also that the two identified approaches (TRA and SRA) were not applied for all the available case studies. The reasons for this lack of information are reported in detail in the previous chapters where each single case study was assessed. However, it is worth to highlight here that in some cases (for example the works carried out in Spain and Morocco), even if the first and the second sampling campaign were completed successfully, the authors presented their results by simply comparing the estimates of erosion rate for the time windows between ^{137}Cs fallout and first sampling and between ^{137}Cs fallout and second sampling, instead of deriving the estimates for the period between the first and second sampling. In these cases, the application of TRA and SRA is still possible and further works on this direction can tell us more about the accuracy of the ^{137}Cs resampling technique in those areas.

The screening of all case studies listed in Chapter 1 (Table 1.1) and especially the validation study presented in Chapter 3 resulted in development of the ^{137}Cs resampling step-by-step procedure which can be presented here as a simple protocol. It comprises of 7 steps (Fig. 4.3):

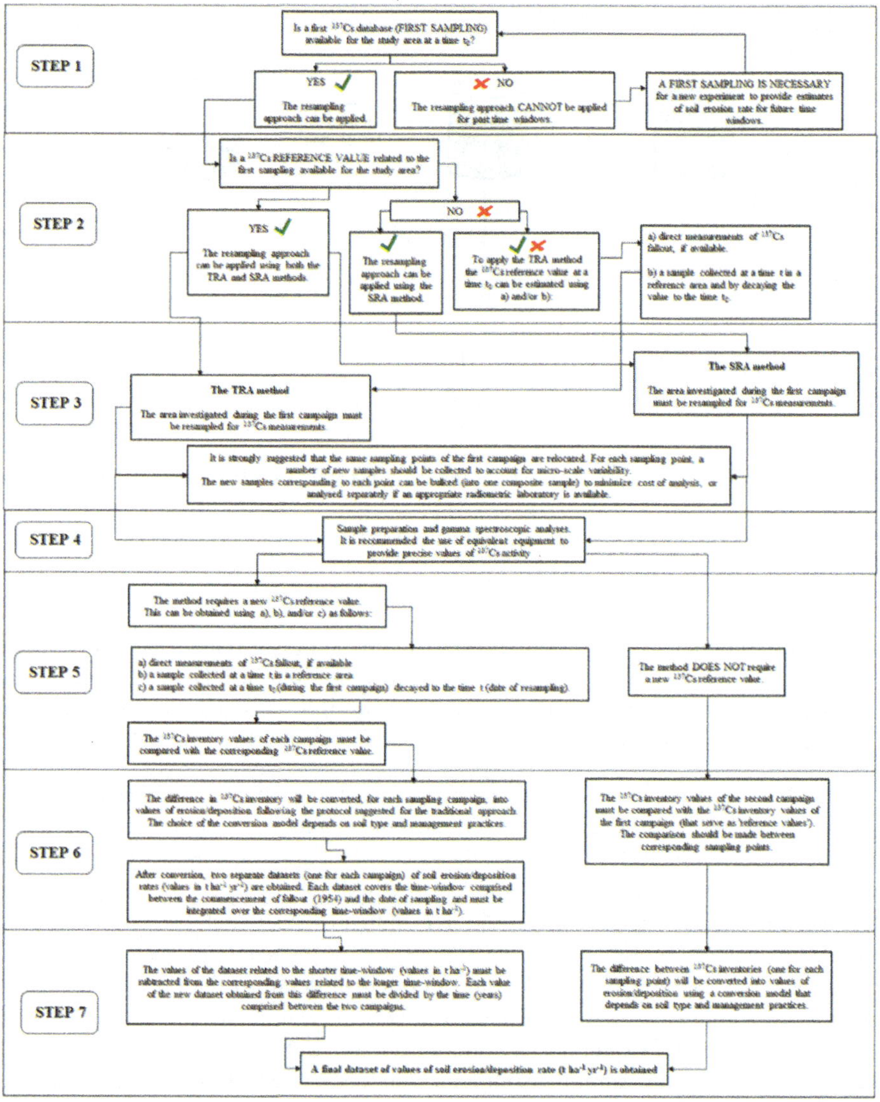

Fig. 4.3 Flow chart describing a step-by-step protocol for the application of the ^{137}Cs resampling approach

1. Selecting and assessing study site and existing ^{137}Cs sampling
2. Selecting the methodical approach (TRA versus SRA)
3. Second sampling campaign of selected study site
4. Sample preparation and gamma spectroscopic analyses
5. Determining the reference value (recalculating the earlier reference value for radioactive decay or new reference sampling)

6. Conversion of ^{137}Cs inventories from second sampling of study site to soil redistribution rates
7. Calculating soil redistribution rates for the time window since the first until the second sampling.

4.2 Possible Study Objectives, Opportunities and Advantages of Using ^{137}Cs Resampling Technique

The successful validation of the plot experiment from Italy described in Chapter 3 and the previous work of Kachanoski (1987) (reviewed in Sect. 2.1) proved that the ^{137}Cs resampling technique is a powerful tool to provide reliable estimates of soil loss for selected time windows. In the light of these results, the two tested approaches (TRA and SRA) can be used to estimate soil erosion rates also in areas where no direct conventional measurements were carried, and this offers great opportunities for future applications.

One of the possible applications of the ^{137}Cs technique is evaluating the effect of land use changes on soil erosion rates. Some of the studies described in this guideline that used the ^{137}Cs resampling approach, showed great potential for assessing the impact of land use changes on soil erosion. The Morocco case study (Sect. 2.3) demonstrated how the ^{137}Cs resampling approach can be used to evaluate the changes in soil erosion rates due to the variation in land use. In that case, a reduction in soil erosion rate was explained by the impact of conservation land management. Similar results were published by Fornes et al. (2005), in the paper described in Sect. 2.1. In that experiment, the reduction of soil loss was attributed to the introduction of grass buffer strips and conservation tillage over a cultivated area in Iowa. Even Loughran and Balog (2006), as it was explained in Sect. 2.1, reported a reduction of soil erosion rates caused by the cessation of clean till cultivation in the Hunter Valley, Australia.

These results are very important from a short-term perspective for land management because they demonstrate the ability of the ^{137}Cs resampling technique to evaluate the effects of different soil conservation strategies in reducing soil erosion. From this point of view, the Chinese case study (Sect. 2.3) is also essential because it demonstrates the beneficial effects of the 'Grain for Green Program' initiated in China in 1999, when some of the cultivated areas of the Loess Plateau were converted into grassland, forest, and shrub land. The reduction of soil erosion rates in those areas proved the effectiveness of this soil conservation program without the need of direct measurements.

Another, important field of application of the ^{137}Cs resampling technique is related to its ability to identify areas affected by soil erosion caused by the impact of climate change. Similar applications require that external disturbance due to possible changes in land use or other anthropic effects are under control. In this respect, the work provided by Spanish case study (Sect. 2.2) demonstrates how the resampling approach can be employed for this purpose. In that study a mountain area of Northern Spain with mosaic of forest and cultivated land was selected and no changes in land use have occurred between the first and the second sampling campaign (timespan of 13 years). This work demonstrated that the changes of soil redistribution rates that occurred in the study area during the study period are associated with a shift in rainfall patterns in the area and the resampling approach proved to be very useful to identify such phenomena. Similar results were obtained by Morocco case study (Sect. 2.4). In that case, the application of the resampling approach in two contrasting sites in Morocco (one agricultural and one with a degraded forest), where no change in land use occurred during the period between the two sampling campaigns (timespan of 15 years), showed an increase of soil erosion rates of 21% and 19%, respectively. The authors explained this increasing trend of soil loss with the impact of extreme weather events occurred in the second period, although further investigations and data are needed to confirm these results. Another attempt to distinguish the effect of climate on soil erosion from the impacts of other dynamic factors was carried out in Southern Italy by Porto et al. (2014). This work, fully described in Sect. 2. 2, involved three experimental catchments in which two field campaigns for ^{137}Cs measurements were undertaken (timespan of 13–15 years). In two of the three catchments investigated, the use of the resampling approach recognised increasing rates of soil erosion in recent times, and this result agrees with other studies that proved evidence of climate change in the area (see Capra et al. 2017; Porto et al. 2022).

Contrasting results were obtained from the forest site in Loess Plateau, China (Sect. 2.3). No change of forest cover has occurred at this study site. The application of the ^{137}Cs resampling approach revealed a decrease in soil erosion rate due to a reduction of rainfall erosivity during the same period, emphasizing the role of climate as a driver of soil erosion in the area.

The studies described above are examples indicating a huge potential of the ^{137}Cs resampling technique for soil erosion investigations and especially for studying the impacts of land use changes, climate, and other erosion factors and for assessment of the efficiency of soil conservation measures. We strongly recommend its application especially in areas where the medium- and long-term strategies and programs for soil conservation, sustainable land management and precision farming are introduced with the perspective of implementation over a time scale of 2–3 decades.

Based on the above premises, even the areas not covered by previous ^{137}Cs measurements can be investigated if the aim is to provide soil erosion rates for future time windows. In other words, if a programme of soil conservation is planned to reduce soil erosion in the next 2–3 decades, or there is expected a change in erosion rates caused by climate change, the resampling approach can be initiated at a new site which was not yet sampled. The first sampling can be done now and the resampling can be planned for the future. In such case the length of investigated time window can be chosen in advance depending on geographical conditions and erosion factors and the expected magnitude of soil erosion. Further, the experiment can be repeated for several successive periods creating a long-term soil erosion monitoring program providing the ^{137}Cs activity in soil is still within the detection limit.

References

Benmansour M, Moussadek M, Moustakim M, Zouagui A, Nouira A, Damnati B, Iaaich H, Mrabet R, Fulajtar E, Mabit L (2019) Use of fallout radionuclides to assess change in soil erosion and sedimentation rates in Northern Morocco. Food and Organisation of the United Nation. Proceeding of Global Symposium on Soil Erosion, 15–17 May 2019, Rome, FAO, pp 656–660

Capra A, Porto P, La Spada, C (2017) Long-term variation of rainfall erosivity in Calabria (Southern Italy). Theoretical and Applied Climatology 128 (1–2): 141–158.

Chappell A, Li Y, Yu HQ, Zhang YZ, Li XY (2015) Cost-effective sampling of ^{137}Cs-derived net soil redistribution: part 2—estimating the spatial mean change over time. J Environ Radioact 144:168–174

de Jong E, Kachanoski RG (1988) The importance of erosion in the carbon balance of prairie soils. Can J Soil Sci 68:111–119

Fornes WL, Whiting PJ, Wilson CG, Matisoff G (2005) Caesium-137-derived erosion rates in an agricultural setting: the effects of model assumptions and management practices. Earth Surf Process Land 30:1181–1189

Kachanoski RG (1987) Comparison of measured soil 137-cesium losses and erosion rates. Can J Soil Sci 67:199–203

Kachanoski RG, de Jong E (1984) Predicting the temporal relationship between cesium-137 and erosion rate. J Environ Qual 13:301–304

Li S, Lobb DA, Kachanoski RG, McConkey BG (2011) Comparing the use of the traditional and repeated-sampling-approach of the Cs-137 technique in soil erosion estimation. Geoderma 160(3–4):324–335

Lobb DA, Kachanoski RG (1999) Modelling tillage erosion in the topographically complex landscapes of southwestern Ontario, Canada. Soil Tillage Res 51:261–277

Loughran RJ, Balog RM (2006) Re-sampling for soil-caesium-137 to assess soil losses after a 19-year interval in a Hunter Valley vineyard, New South Wales, Australia. Geogr Res 44(1):77–86

McHenry JR, Bubenzer GD (1985) Field erosion estimated from ^{137}Cs activity measurements. Trans Am Soc Agric Eng 28:480–483

Moustakim M, Benmansour M, Zouagui A, Nouira A, Benkdad A, Damnati B (2019) Use of caesium-137 re-sampling and excess lead-210 techniques to assess changes in soil redistribution rates within an agricultural field in Nakhla watershed. J Afr Earth Sc 156:158–167

Porto P, Walling DE, Alewell C, Callegari G, Mabit L, Mallimo N, Meusburger K, Zehringer M (2014) Use of a ^{137}Cs re-sampling technique to investigate temporal changes in soil erosion and sediment mobilisation for a small forested catchment in southern Italy. J Environ Radioact 138:137–148

Porto P, Walling DE, Callegari G (2018) Using repeated ^{137}Cs and ^{210}Pbex measurements to establish sediment budgets for different time windows and explore the effect of connectivity on soil erosion rates in a small experimental catchment in Southern Italy. Land Degrad Dev 29:1819–1832

Porto P, Bacchi M, Preiti G, Romeo M, Monti M (2022) Combining plot measurements and a calibrated RUSLE model to investigate recent changes in soil erosion in upland areas in Southern Italy. J Soils Sediments 22:1010–1022

Tiessen KHD, Li S, Lobb DA, Mehuys GR, Rees HW, Chow TL (2009) Using repeated measurements of ^{137}Cs and modelling to identify spatial patterns of tillage and water erosion within potato production in Atlantic Canada. Geoderma 153(1–2):104–118

The manufacturer's authorised representative in the EU is Springer Nature Customer Service Centre GmbH, Europaplatz 3, 69115 Heidelberg, Germany. If you have any concerns regarding our products, please contact ProductSafety@springernature.com

Printed and bound by CPI Group (UK) Ltd, Croydon, CR0 4YY
01/10/2026
02225916-0008